Spaces of Congestion and Traffic

This book provides a political history of urban traffic congestion in the twentieth century, and explores how and why experts from a range of professional disciplines have attempted to solve what they have called 'the traffic problem'.

It draws on case studies of historical traffic projects in London to trace the relationship among technologies, infrastructures, politics, and power on the capital's congested streets. From the visions of urban planners to the concrete realities of engineers, and from the demands of traffic cops and economists to the new world of electronic surveillance, the book examines the political tensions embedded in the streets of our world cities. It also reveals the hand of capital in our traffic landscape.

This book challenges conventional wisdom on urban traffic congestion, deploying a broad array of historical and material sources to tell a powerful account of how our cities work and why traffic remains such a problem. It is a welcome addition to literature on histories and geographies of urban mobility and will appeal to students and researchers in the fields of urban history, transport studies, historical geography, planning history, and the history of technology.

David Rooney has held curatorial positions at both the Science Museum and the National Maritime Museum. Over a 22-year career he has contributed to several critically acclaimed exhibitions and galleries, including the RIBA-award-winning *Mathematics: The Winton Gallery*, designed by Zaha Hadid Architects, for which he was lead curator. In 2016 he obtained his PhD from Royal Holloway, University of London, where he is now an Honorary Research Associate.

Transport and Mobility

Series Editor: John Nelson

The inception of this series marks a major resurgence of geographical research into transport and mobility. Reflecting the dynamic relationships between socio-spatial behaviour and change, it acts as a forum for cutting-edge research into transport and mobility, and for innovative and decisive debates on the formulation and repercussions of transport policy making.

Intermodal Freight Terminals
A Life Cycle Governance Framework
Jason Monios and Rickard Bergqvist

Railway Deregulation in Sweden
Dismantling a Monopoly
Gunnar Alexandersson and Staffan Hulten

Community-Owned Transport
Leigh Glover

Geographies of Transport and Mobility
Prospects and Challenges in an Age of Climate Change
Stewart Barr, Jan Prillwitz, Tim Ryley, and Gareth Shaw

Urban Mobilities in the Global South
Edited by Tanu Priya Uteng and Karen Lucas

Spaces of Congestion and Traffic
Politics and Technologies in Twentieth-Century London
David Rooney

For more information about this series, please visit: www.routledge.com/Transport-and-Mobility/book-series/ASHSER-1188

Spaces of Congestion and Traffic

Politics and Technologies in Twentieth-Century London

David Rooney

LONDON AND NEW YORK

First published 2019
by Routledge
2 Park Square, Milton Park, Abingdon, Oxon OX14 4RN

and by Routledge
52 Vanderbilt Avenue, New York, NY 10017

First issued in paperback 2020

Routledge is an imprint of the Taylor & Francis Group, an informa business

British Library Cataloguing-in-Publication Data
A catalogue record for this book is available from the British Library

Library of Congress Cataloging-in-Publication Data
A catalog record has been requested for this book

ISBN 13: 978-0-36-758790-1 (pbk)
ISBN 13: 978-1-138-58073-2 (hbk)

Typeset in Times New Roman
by Integra Software Services Pvt. Ltd.

Contents

Figures

Tables

Preface

The geography department at Royal Holloway, University of London, where I carried out the studies that led to this book, was the most stimulating, creative, supportive, inspirational, thought-provoking, and friendly institution I could have imagined working in. I joined ready for a challenge and that was what I got, but it was an utterly delightful one. It transformed me intellectually, but also with a new emotional connection with the world, a substantially more critical moral framework, and a physical restlessness to explore the subject of my studies – the streets of London. All this from a simple course of study.

For this I must express my most sincere thanks to the staff and students of the department. They all, together, constituted this remarkable oxygenated atmosphere. They are too many to name, but let me pick out three for special mention. Mustafa Dikeç was my doctoral adviser for the first three years or so and I greatly valued our conversations on many topics. On Mustafa's departure from Royal Holloway, Peter Adey took over the advisory role and was generous with his time and knowledge, for which I thank him enormously. But my most fervent thanks and acknowledgement must go to my supervisor, David Gilbert, who was a constant, never-failing guide throughout this expedition, and I appreciated every word, gesture, and moment of his calm and wise counsel. I also wish to record my appreciation to Royal Holloway for the College Research Scholarship that funded my study. Simon Gunn and Michael Hebbert examined my thesis and have been hugely supportive, for which I thank them.

Early ideas underpinning aspects of this book were presented in November 2010 at the *Blocked Arteries: Circulation and Congestion in History* conference at the Institute of Historical Research (IHR), London, organized by Carlos Lopez Galviz and Dhan Zunino Singh. This conference was a formative experience for me and I appreciated the lively and supportive discussion therein. I enjoyed and benefited greatly from many subsequent conversations with Carlos on other projects. My presentation at the IHR led to a paper published as 'Visualization, Decentralization and Metropolitan Improvement: "Light-and-Air" and London County Council Photographs, 1899–1908', *Urban History* 40, no. 3 (August 2013): 462–82. This enabled me to think through ideas about modernity, representation, and the politics of urban mobility that have been most useful as this project has developed, and I am grateful to the editors and referees for giving me this early boost. Parts of Chapter 4 were presented in July 2015 at the International Conference of Historical Geographers at the Royal Geographical Society, London, on a panel organized by Richard Dennis, Deryck Holdsworth, and Phillip Mackintosh. I appreciated the opportunity to discuss matters with such a wise group of scholars, and subsequently to publish a version of my paper in the 2018 Routledge volume *Architectures of Hurry: Mobilities, Cities and Modernity*. Chapter 5 was published in a modified form in 2014 in *Twentieth Century British History*. My thanks to the editors and referees for their generous and supportive comments. Parts of Chapter 6 were presented in March 2015 at the *SPUD IV* meeting at Lincoln College, Oxford, organized by Simon Gunn and Otto Saumarez-Smith. I am most grateful to Simon and Otto for their kind invitation and to all those at the meeting for their friendly and energizing conversation. Parts of Chapter 7 were presented in October 2015 at Royal Holloway's *Landscape Surgery* convened by Veronica della Dora. This came at a particularly crucial time in the refinement of my thoughts and I found the discussion hugely valuable.

I have benefited from countless conversations with people not directly connected with my research area, in particular James Nye (with whom I have long been researching and writing on matters involving other aspects of urban infrastructure) and James Naylor, a senior consultant at McKinsey, who has read drafts and made helpful observations, many from left field.

At the Science Museum, my employer for the duration of this project, everybody has been patient and supportive over the years. A few deserve special thanks. Tim Boon, Head of Research and my manager in 2010, accepted with vigour my request to carry out a PhD project. I valued his unfailing support in my putting an application together, and subsequently throughout the project. Heather Mayfield, then the museum's Deputy Director, agreed immediately to my request and offered kind words whenever I needed a boost. Jean Franczyk, who succeeded Heather, took an equally supportive interest. John Liffen, Robert Bud, and Peter Morris, similarly, took part in numerous creative discussions over where my research might take me. Tilly Blyth became my manager partway through the project and backed me all the way since. It is hard to study and write when the day job is demanding, and Tilly helped make it possible by shielding me from things that could wait. Her words of advice and wisdom steeled me many times, for which I thank her. Heading our department was Hadrian Ellory-van Dekker, who was equally supportive and understanding when the pressures of studying while working occasionally bubbled up to the surface. On Hadrian's departure, Tilly took his role and continued to give her support. Finally, Andrew Nahum, now Keeper Emeritus but then Senior Keeper, has been a close friend as well as valued colleague, mentor, sounding-board, and guide since I first worked for him in the 1990s on a big gallery project. Andrew taught me (among other things) how to sniff out a good story, and I think I've found a few here.

Finally, I wish to thank the librarians, archivists, curators, and other specialists for the help they have given me during this project.

Attempts have been made to trace all copyright holders and to obtain their permission for the use of copyright images. I apologize for any errors or omissions and would be grateful if notified, via the publishers, of any corrections that should be incorporated in future editions of the book.

This book contains public sector information licensed under the Open Government Licence v3.0. See www.nationalarchives.gov.uk/doc/open-government-licence/version/3/.

This work is dedicated with love to my family.

David Rooney
London, March 2018

1   Introducing the traffic problem

Introduction

'Vehicle speeds in Central London are falling', wrote Ruth Bashall and Gavin
Smith in a 1992 account of London's transport crisis, 'and at 10 mph are
currently little faster than the horse and cart of the turn of the century'.[1] This
is a common refrain and, as will be shown presently, an accurate one. But it
does not do much to scratch the surface of the *experience* of traffic in
London, which depends on the person doing the experiencing, and varies
through time and from place to place. Perhaps that is the point of singular
statistics: they imprison the breadth of experience in a black box and hide it
from view, so that the only reasonable response can be one of dismay at how
bad things have got – at the crisis we find ourselves in. No faster than a
horse and cart? So much for progress. Yet, for a crisis situation, it seems to
have been with us for a long time. In 1938, London's *Evening Standard*
newspaper carried out a publicity stunt focusing attention on the capital's
traffic jams whereby it drove a van fitted with a large four-sided clock on
timed journeys across London, demonstrating where and why congestion
occurred.[2] During the stunt, both the British Road Federation and the House
of Lords lamented that the newspaper van did not travel any faster than a
horse.[3]

The *Evening Standard*'s timed trips followed a project in summer 1936, in
which the Ministry of Transport carried out, for the first time, systematic research
into traffic speeds and delays caused by congestion in the capital. The newspaper
stunt might well have been inspired by the Ministry research project, in which an
Austin Light-Six motor-car, driven by a professional chauffeur, ferried officials
with stop-watches and clipboards on a series of journeys across four routes
through the capital.[4] The first route ran 12.6 miles from Chiswick in the west to
Bow Road in the east. The second ran the same length from Hornsey in the north
to Streatham in the south. The third route ran from Golders Green in the north-
west to Lewisham in the south-east. The fourth was a route of 22.75 miles from
Chiswick around the North Circular Road to Ilford. Journey times were recorded,
as were the locations and duration of delays and stoppages. The driver was
'steady and competent', with 'no inducement to attempt to break records or to

take risks'. He therefore represented the 'punctiliously cautious and considerate driver who presumably constitutes the bulk of the British motoring community'.[5]

After the officials had spent several weeks plying the routes daily, the results showed an overall average speed across the three cross-London routes of 12.5 mph. But it was the west-to-east route through the City that was the slowest, averaging 5.85 mph with the worst journey averaging just 3.6 mph. These journeys were 'painfully slow' and 'ceaselessly congested', according to the official report.[6] Next-worst was the route from Euston Road south to Trafalgar Square, with the slowest journey averaging 6.3 mph. The problems occurred most markedly at junctions, and the report listed the most problematic intersections, including Ludgate Circus, Bank, Gardiner's Corner at Aldgate, St Giles' Circus, and the junction of Euston Road and Tottenham Court Road. By contrast, the circular route avoiding Central London was much faster, with average speeds of 23.6 mph, meaning it was often quicker to go the long way around. Commentators, noting the Ministry of Transport's 1936 traffic survey work, asked 'Is a road crisis developing?'[7]

This book examines traffic congestion in twentieth-century London, focusing largely on the period to about 1980, but with some excursions into more recent territory. It surveys the ways congestion has been considered in the history of urban planning, and examines a range of alternative 'solutions' to the problem as well as how they have been negotiated into reality. In doing so, it will decode 'the traffic problem', setting it into wider geographical, political, and technological contexts.

The traffic problem in history

The answer to the question posed in 1936 about a road crisis seems obvious. It is all we talk about when we discuss transport in the capital – the congestion, the fact that we go no faster today than in the age of the horse. It is all we have *ever* talked about, as London's canon of modern-day chroniclers has described. Peter Ackroyd, for instance, tracks complaints about the traffic problem back 500 years, noting that 'The state of traffic in the capital was a source of constant complaint in the sixteenth century, as it has become for each generation.'[8] Stephen Inwood notes that there was a brief improvement following the seventeenth-century Great Fire, but that 'in the eighteenth century the traffic problem grew worse again' and, the following century, 'London's traffic congestion went from bad to worse.'[9] Calls for something to be done became more insistent in Regency times, as James Winter has described, noting 'officialdom's growing concern about the traffic problem' in the 1830s, when new traffic legislation expressed the seriousness with which it was taken.[10] Roy Porter told a similar tale of traffic woe in the nineteenth century, a time when 'London's traffic problems were becoming ominous' and 'jams could be grim'.[11] As the nineteenth century gave way to the twentieth and the motor-car joined the streetscape, the problem just continued to get worse. Jonathan Schneer observes that, in 1900, 'London traffic jams were notorious, new-fangled horseless carriages and traditional horse-drawn vehicles often merging in near gridlock conditions.'[12] Jerry White agrees, commenting that 'Traffic was one of the enduring problems of the nineteenth century ... and so it remained throughout the twentieth.'[13] The traffic problem is a refrain, ever present in the mouths of Londoners and London historians. We keep returning to it when we speak of London.

Amid this constant background of general complaint can be discerned peaks of concern. Two such peaks will be considered time and again in this study. The first spanned the late 1920s and early 1930s. Joe Moran has noted that this was a period of critical importance in the traffic experience, with the foundation of the Pedestrians' Association in the face of growing concerns over road safety, a Road Traffic Act that sought to impose responsibilities on motorists, and the publication of the Highway Code.[14] The Ministry of Transport, with its 1936 traffic survey, was thus responding to a growing concern about traffic. A second peak of concern over London's traffic was the period from the late 1950s to the early 1960s, this time over what Simon Gunn has described as a 'motor revolution' in the expansion of automobility in urban Britain, when motor transport was 'high on the political agenda'.[15]

One example demonstrates the seriousness with which 1930s commentators viewed the traffic problem. In 1933, a former railway worker and transport writer, Henry Watson, published the book *Street Traffic Flow*. It was a major study of urban traffic congestion, and in order to treat the traffic system as a whole, Watson consulted a wide range of bodies, including the Ministry of Transport, Home Office, Automobile Association, Society of Motor Manufacturers and Traders, National Safety First Association, Pedestrians' Association, National Horse Association, Metropolitan Police, tram and bus companies, journal editors, and traffic signal manufacturers, as well as traffic specialists in the USA.[16]

Watson's study showed the traffic problem in fine grain. He differentiated between traffic types and their speeds and handling characteristics. He considered the way traffic varied from city to city, between residential and industrial districts, by time of day, by season, and by weather. He also observed the irregular, minute-by-minute variation of flow in busy streets, as well as the lateral position adopted by different vehicle classes at a time when camber mattered, surfaces were often slippery with oil and horse manure, and many streets included tramway tracks. He provided exhaustive data on delays owing to obstructions; analysis of flow over different types of junction; accident statistics; and the effects of traffic signals, of pedestrian crossings, and of parked vehicles. He included 21 newly commissioned photographs representing traffic problems in London's streets; 89 graphs, illustrations and diagrams; 35 tables of statistics; and an extensive bibliography. He concluded his account with an assessment of the economics and politics of traffic, considering how time could be given value in order to estimate the costs of congestion, and whether certain users should be subject to restrictions.

Watson's book gives us a clear sense of the technical problems of traffic. Yet it also shows that apparently even-handed representations contain political bias.

Watson favoured tram travel, and the book comes down strongly in its favour against motor buses. He claimed that:

> For heavy passenger transport the bus is commonly – and in London always has been – an overrated proposition, appealing to those impressed by the obvious and exceptional, and lacking the perspective and technical knowledge to appreciate their shortcomings as compared with tramways.[17]

His book was published in the same year that the London Passenger Transport Bill was enacted, bringing London's transport providers under public control and sounding the death knell of the tram network in favour of an aggressive expansion of trolley- and motor-bus travel. Watson wrote amid bitter arguments between tram and bus promoters that were really an ideological fight between municipal socialism and free-market capitalism.[18] *Street Traffic Flow* is an underappreciated gem that acted as propaganda in a fierce war for control of London's political economy.

An example of 1960s public concern over traffic came with the Oscar-nominated film *Automania 2000*, made in 1963 by the acclaimed animators John Halas and Joy Batchelor, and described by the British Film Institute's animation curator as one of the greatest animated films of all time.[19] It offered a dystopian vision of a world in the year 2000 submerged by motor cars, fuelled by demand for cheap consumer goods made possible by 'scientists', the villains of the piece. Before long, so many cars had been manufactured that it had become impossible to move, and the population had been forced to adapt to living *in* their cars, piled high in the streets in a condition of 'universal immobility'.

The narrator intoned: 'In the more densely populated areas of our planet, people have been confined to their cars for over five years. The younger children cannot remember the time when it was possible to move around in cars.' The final act of the scientist, before he was killed by one of his own creations, was to invent sentient cars that reproduced themselves. The car had finally taken over humanity.

Automania 2000, like *Street Traffic Flow* thirty years previously, was a political statement rather than a dispassionate assessment of an uncontested traffic problem. Halas and Batchelor were commenting on the hubris of the technocrat, so greatly lauded in the white heat of 1960s Britain. For them, the problem was people, not traffic. That was Watson's problem too – public transport was a proxy for wider political battles over resource allocation in liberal economies. Traffic stands for other things. Roads are political and economic spaces as well as geographical and physical places, and the solutions proposed depend on the problem one sees – and on the world view one holds. Congestion is not a stable concept. Talk of the traffic problem encodes wider concerns.

The intention of this book is thus to decode the traffic problem, deconstructing it and rethinking its nature, both to understand traffic in London better and to shed light on ways in which physical urban infrastructures such as London's streets have been mobilized in political, social, and economic discourses.

It will do this by considering congestion as a complex network that is sociological and political as well as spatial, human, and technological. This will involve looking at a wide variety of actors and the interactions between them, where actors are not merely people or technologies but socio-technical nodes. An examination of the politics of congestion will demonstrate that traffic has different values – economic, aesthetic, professional, moral – to different people. Three questions emerge from this.

Firstly, what have been the dominant characterizations of the traffic problem, and what alternative characterizations have been obscured by the dominant framings? In the popular discourse, the traffic problem is congestion, with average traffic speed in London held up as the emblematic statistic of a pathological condition. This has been translated in the professional discourse as a failure of urban planning, with a failure of governance intimately related to it. In the scholarly discourse, the dominant characterization of this failure of planning has until recently been based on ideas that plans are singular, holistic, stable, and have clear authorship, although new scholarship is revising this view. Thus, taking this all into account, the standard story is that the traffic problem would be solved if congestion was reduced and traffic moved faster; this would be enabled by better governance, allowing planners to get on with their work; and if only Wren and Abercrombie's plans for London had been enacted, the whole problem would have been sorted long ago. We're paying the price today, we are often told.

Of course, this is a gross oversimplification of the way traffic has been characterized, but it points up the clear value in problematizing dominant accounts and looking to their margins. This book looks at plans as unstable artefacts with multiple authors and influences. It widens the scope of people who might be termed 'planners'. It casts doubt on the ability of traffic statistics to represent reality. By considering alternative characterizations of the traffic problem, such as market failure and system failure, it looks beyond professional planning and hard road building to consider softer (though no less physical) interventions such as road pricing and traffic lights. And by reframing traffic as a system of movement it is possible to shift from 'congestion' to 'friction' as a means to characterize busy streets.

The second question to emerge is what the relationships between London's traffic infrastructure and capital are. The relationships between London *property* and capital are manifold and have been well considered. Less common is to consider the capital's street infrastructure – the built environment *between* the highly capitalized buildings – as part of the same account. This research question is therefore an attempt to place traffic into a wider account of capitalism and the state, specifically the political economies of markets. Streets, like buildings, are saturated with market decisions. If Ken Livingstone's 2003 Congestion Charge scheme is a poster-child for modern solutions to congestion, and road pricing a refrain heard as often as the need for new roads, what can its history tell us about the Keynesian consensus, neoliberalism, Thatcherism, social democracy, and the

New Labour turn? Where does traffic sit with gentrification? What were the relationships between property developers and council planners in the shaping of London's traffic scene? This is not a straightforward linear story nor a wholly recent one, and one conclusion is that the traffic problem has in part been constructed and mobilized in the service of accumulating capital and effecting political change, albeit in an ambivalent way. Another conclusion, however, is that capital is just a necessary, not a sufficient, explanation for the experience of mobilities in London.

The third question asks what is distinctively 'London' about London's traffic problem, and what the relationship is between London and other places. This question seeks to understand the traffic problem on a variety of geographical scales from global to local, as well as understanding how local relates to global and vice versa. In one sense, this is a question of geography, or rather how London's topography and form might dictate its traffic problem. It does not, for instance, have a gridded street plan. How does this matter? In another sense it is a technological question, or one of how technologies travel. But it is most significantly a socio-political question. In answering it, the book will examine, for instance, how sociological ideas such as racial segregation in the USA, Asia, and Africa relate to the concept of pedestrian segregation in the City of London's pedestrian walkway network or the guardrails along the edges of east London pavements. This type of narrative is a logical outcome from an approach informed by ideas of networks. The evidence will suggest that some aspects of the London story, while wholly connected to a global network, are unique to the capital, while in other ways the case of London can be used as a model for thinking about complex urban problems in other cities.

The traffic problem in statistics

Numerous official bodies have recorded traffic statistics of one form or another, and at one time or another. These have included the Board of Trade, the London and Home Counties Traffic Advisory Committee, Royal Commissions, Select Committees, the London County Council (LCC), and the Greater London Council (GLC). Today, both the Department for Transport (DfT) and Transport for London (TfL) publish detailed traffic statistics, but even DfT observes that 'Traffic congestion is an inherently difficult concept to define as it has both *physical* and *relative* dimensions ... congestion can mean very different things to different people.'[20] It is hard, therefore, to decide what to survey when seeking to measure traffic and congestion: journey time or average speed, quantity of traffic or number of accidents. The statistics presented reflected the picture that authors wished to paint. Henry Watson, for instance, who wrote *Street Traffic Flow* in 1933, provided the figures shown in Table 1.1. However, he chose not to present the vehicle classes in order of magnitude or percentage change. Instead, it served his case to show that buses increased more than trams. As was shown earlier, Watson spoke for the tram lobby at a time of febrile argument over municipal socialism in Britain's cities.

Table 1.1 Census of London traffic in busy areas, 1912–1926.

Census of London traffic: thousands of vehicles passing 39 selected points

	1912	1920	1923	1926	*% increase since 1912*
Motors	253	364	456	587	132
Buses	129	129	166	194	50
Trams	26.9	27.3	30.4	31.8	18
Horse vehicles	307	182	133	105	−66
Cycles	65.8	91.6	131	188	185
Barrows etc.	44.7	24.5	25.3	22.0	−51
Total	**825**	**818**	**943**	**1129**	**37**

Source: Henry Watson, *Street Traffic Flow* (London: Chapman and Hall, 1933), 4.

Watson drew his data in part from annual statistics published by the LCC and GLC, which were gathered from the LCC's formation in 1889 to the end of the GLC in the 1980s (with a gap from 1938 to 1945). These mostly favoured traffic counts (as Watson showed), which are hard to translate into experiences, but give a sense of the geography of the problem. In 1938, a comparative survey was published (shown in Table 1.2) showing the location of some of London's busiest traffic locations and the rise in traffic density since 1904.[21] It is clear that traffic levels had risen greatly since the start of the century, in many cases almost tripling and in some instances rising even faster, such as at Shepherd's Bush Green, where traffic in 1938 was almost five times that of 1904.

Census data help reveal the *locations* of London's traffic problems but do little to reveal the *effects* of congestion. For this, average speed has become the common characterization of the user experience, as has been noted. A recent travel report from TfL included average traffic speeds for 2013, which are shown in Table 1.3.[22] These may be compared with figures from 1906, when the LCC published a table listing average speeds of different vehicles in London's central area, shown in Table 1.4.[23] Private motor-cars were still not common on London's streets in 1906, but we can take the figures for motor-cabs as reasonably representative of motor-cars in general (although cabs spent time crawling for fares, which will have brought the average down slightly). These figures are useful in that they were captured at a time when motorized vehicles were still in the minority, and give some evidential weight to claims that average traffic speeds in the capital have not increased since the horse-drawn era.

That was the situation at the start and end of the period being considered. Statistics exist for much of the interim too. We recall from the beginning of the chapter that the average speed recorded in 1936 in the central area was 12.5 mph, and the rest of the situation between 1906 and 2013 can be glimpsed by examining firstly a series of average speed statistics collected by the Road Research Laboratory, the Ministry of Transport, and the GLC over the period from 1952 to 1968, which did not

Table 1.2 Traffic census data at specific locations, 1904–1937.

Total number of vehicles enumerated during each census (8 a.m. to 8 p.m.) at a number of important points on a day in July, 1904–1937.

The points are arranged in order of magnitude in the latest year.
The weather conditions at the date of each census (except 1904 and 1915) were as under –

1912 Raining all day.	1919 Showery.
1927 Generally fine.	1931 Dull, rain 5 to 8 p.m.
1935 Generally fine.	1937 Generally fine.
	1923 Generally fine.
	1933 Showery, bright intervals.

Point of enumeration	*1904*	*1912*	*1915*	*1919*	*1923*	*1927*	*1931*	*1933*	*1935*	*1937*
Hyde-park-corner	29,286	41,106	35,342	36,773	56,039	a 63,005	a 69,011	a 81,857	a 82,728	a 80,536
Trafalgar-square	27,768	34,897	34,561	31,071	42,042	a 57,652	a 62,542	a 64,735	a 66,039	a 65,406
Marble-arch	29,320	31,927	25,794	25,612	35,594	a 45,989	a 38,129	a 52,197	a 60,090	a 57,698
Piccadilly-circus	27,050	39,322	32,911	32,080	41,270	a 43,210	a 47,823	a 51,160	a 48,171	a 49,888
Blackfriars-bridge-approach (Chatham-place)	24,385	30,191	—	24,391	29,611	37,053	40,491	41,740	49,395	48,514
Piccadilly at St. James's-st.	20,474	c 16,185	21,717	22,705	31,894	33,070	39,394	42,839	44,035	47,414
Edgware-road at Praed-street and Marylebone-road	—		—	c 16,859	cd 21,250	c 25,173	36,260	45,028	45,190	44,112
Shepherds Bush-green	9,202	9,849	—	12,940	18,314	30,167	33,950	37,279	40,612	43,705
Ludgate-circus	22,956	27,949	—	18,659	24,361	31,594	34,075	35,473	39,881	42,434
High Holborn at Southampton-row	e 14,590	24,286	21,472	18,984	26,611	32,189	37,742	34,956	37,240	39,411
Knightsbridge at Sloane-street	21,471	28,620	f 19,384	22,215	31,031	33,172	34,711	34,411	39,191	39,396

Mansion-house	27,523	30,934	—	22,775	31,120	36,184	35,216	35,784	38,985	38,640
Hammersmith-broadway	11,587	12,974	—	13,785	22,792	31,379	b 29,130	34,852	37,683	38,377
Oxford-circus	24,228	27,310	21,303	23,683	31,289	32,068	36,513	37,459	38,949	36,781
Elephant and Castle	16,176	20,782	18,856	17,716	26,329	31,683	32,499	32,648	34,859	36,420
Vauxhall-cross	—	19,035	16,537	15,881	23,091	28,556	30,933	32,213	38,245	35,391
St. Giles-circus	21,336	24,479	21,700	18,290	26,285	29,005	28,742	32,713	34,226	34,591
King's-cross	—	13,493	—	12,901	23,965	27,140	*28,£34	28980,	31,228	34,547
Monument	—	24,344	—	19,206	25,550	30,722	33,846	32,126	31,645	32,966
Blackfriars-rd. at Stamford-st.	17,214	18,980	16,581	14,953	19,554	23,738	25,255	26,107	30,575	32,444
Strand at Wellington-street	19,743	24,743	—	19,698	d 24,578	33,918	36,662	37,223	36,211	h 31,864
Kensington High-street at Church-street	10,900	15,989	11,758	14,562	20,973	25,111	26,722	29,567	30,588	31,536
Holborn-circus	20,906	24,412	—	16,771	21,655	25,760	25,764	26,036	29,922	31,212
Euston-road at Tottenham-court-road	—	6,308	—	14,138	21,847	24,204	b 20,767	27,163	29,297	29,794
'Angel', Islington	—	13,606	—	11,750	20,201	24,473	25,168	26,291	27,651	29,743
City-road at Old-street	—	16,002	14,181	13,167	19,697	22,620	23,957	23,059	26,740	29,117
Queen Victoria-st. at Cannon-street	20,108	24,131	—	19,205	22,786	28,777	29,438	30,099	30,796	28,556
Old-street at Kingsland-road	—	12,255	—	6,897	15,313	18,928	20,692	22,261	26,384	28,297
Gray's Inn-rd. at Theobald's-rd.	—	17,822	17,419	13,987	20,818	21,348	23,165	24,253	26,470	27,485
Cheapside at Newgate-street	19,651	20,213	—	14,153	17,802	23,191	25,581	25,400	27,985	27,320
Notting-hill-gate	12,297	13,049	—	9,982	17,158	23,612	26,158	27,942	27,318	27,195
London-bridge, South-side	—	15,850	17,961	13,001	18,662	23,493	24,140	22,453	24,061	26,025
Bishopsgate at Liverpool-street	—	18,295	—	14,591	16,705	21,937	22,051	21,793	24,002	26,024

(Continued)

Table 1.2 (Continued)

Point of enumeration	1904	1912	1915	1919	1923	1927	1931	1933	1935	1937
Cheapside at King-street	14,190	17,313	—	11,311	15,470	19,843	19,330	21,643	23,027	24,246
Whitechapel High-street, Gardiner's-corner	17,932	19,498	17,775	17,275	18,448	23,169	27,086	28,130	28,800	b 22,480
Aldgate at Minories	14,205	13,425	g 13,209	13,009	15,421	18,185	20,563	20,510	22,739	21,873
Cornhill at Bishopsgate	11,738	12,648	—	11,256	14,425	18,378	19,902	19,579	21,686	21,484
Moorgate at London-wall	—	16,784	—	8,650	13,998	16,991	17,779	18,194	20,549	20,747

(a) Roundabout or one-way system of traffic control in operation.
(b) The census was affected by road repairs.
(c) Prior to 1928 the census was confined to the junction of Edgware-road with Praed-street.
(d) The 1923 census at Edgware-road junction with Praed-street and Strand at Wellington-street were [sic] affected by an upheaval of the roadway following heavy rain.
(e) The 1904 census was taken at High Holborn junction with Red Lion-street.
(f) The 1915 census was taken at the junction with Albert Gate and not at the junction with Sloane-street.
(g) The 1915 census was taken at the junction of Aldgate High-street with Mansell-street.
(h) General traffic was not permitted to go northwards over the temporary Waterloo-bridge.
* Error is in source.

Source: London County Council, *London Statistics 1936–38* (London: London County Council, 1938), 317.

Table 1.3 Average traffic speeds on working weekdays, 2013.

	Morning-peak speed (mph)	Inter-peak speed (mph)	Evening-peak speed (mph)
Central area	9.38	8.45	8.57
Inner London	12.4	13.2	11.2
Outer London	19.5	21.8	18.1

Source: Transport for London, *Travel in London: Report 7* (London: Transport for London, 2014), 145 (converted from kph to mph).

Table 1.4 Speed of vehicles in the central area of London, *c.* 1906.

Class of vehicle	Speed during 'crush-hours' (8–10 a.m. and 5–7 p.m.) (mph)	Speed during slack hours (mph)
Traction engines	2	2
Heavy vans	2½	3
Light vans	4 to 6	6 to 8
Omnibuses – horse	3½ to 6	5¼ to 8
Omnibuses – motor	6½ to 8¼	8¼ to 11½
Cabs – horse	3½ to 6	6 to 8½
Cabs – motor	8	12
Tramways – horse	2 to 5	5½ to 8
Tramways – electric	5½ to 7	8½ to 11½

Source: London County Council, *London Statistics 1905–6* (London: London County Council, 1906), 334.

Table 1.5 Average traffic speed in central London, 1952–1968.

	1952	1954	1956	1958	1960	1962	1964	1966	1968
Off-peak speed (mph)	11.1	10.8	10.3	10.0	9.7	10.3	10.6	10.7	11.2
Evening-peak speed (mph)	10.9	9.9	9.1	8.3	8.6	9.5	8.7	9.5	11.1

Source: Greater London Council, *Annual Abstract of Greater London Statistics 1968* (London: Greater London Council, 1968), 61.

differentiate for different vehicle types but, instead, like the 1936 experiment, showed the average speed of a survey car. They are shown in Table 1.5.[24]

Secondly, in 1978, the GLC was able to paint a slightly rosier picture by adjusting the scope of the averaging. Here, the average speed in the central area remained about the same as that of the 1960s, but speeds on the GLC 'primary roads' trunk network and those on roads in outer areas were higher. By including them in the averaging, the council could claim that average speeds across all roads topped 20 mph (see Table 1.6).[25]

Finally, in 1998, the Department for the Environment, Transport, and the Regions (DETR) published a historical data series filling in the period from 1968 to 1998, but only for the central and inner areas, as shown in Table 1.7.[26]

Table 1.6 Average traffic speeds in Greater London, various dates from 1975 to 1978.

	Morning-peak (mph)	Off-peak (mph)	Evening-peak (mph)
Primary roads	24.9	37.1	30.4
Central area	12.3	12.6	11.9
Inner area	13.3	16.0	12.9
Central + inner	13.0	13.0	12.5
Outer SE	16.9	24.1	18.7
Outer SW	19.8	22.2	19.4
Outer NW	18.5	23.9	18.2
Outer N	17.5	22.9	19.8
Outer NE	17.9	23.7	18.4
All outer areas	18.3	23.3	18.8
All areas	16.4	19.3	16.2
All roads	**17.4**	**21.1**	**17.6**

Source: Greater London Council, *Annual Abstract of Greater London Statistics 1979 and 1980* (London: Greater London Council, 1980), 88.

Table 1.7 Average traffic speeds (mph) in inner London, 1968–1998.

	1968	1971	1974	1977	1980	1983	1986	1990	1994	1997
Central area morning-peak speed	12.7	12.9	14.2	12.3	12.1	11.8	11.5	10.3	10.9	10.0
Central area inter-peak speed	12.1	12.6	12.9	12.6	11.6	11.9	11.0	10.6	10.9	10.0
Central area evening-peak speed	11.8	12.7	13.2	11.9	12.2	11.5	11.0	10.3	10.8	10.2
Inner area morning-peak speed	15.1	14.5	15.9	13.9	14.2	13.5	11.8	13.3	13.4	12.0
Inner area inter-peak speed	18.3	18.6	18.6	17.3	17.2	16.3	14.6	15.8	15.0	14.8
Inner area evening-peak speed	15.2	14.5	15.5	13.5	14.1	13.1	11.6	13.2	12.8	11.4
Central + inner area morning-peak speed	14.4	14.1	15.4	13.5	13.6	13.0	11.7	12.3	12.6	11.4
Central + inner area inter-peak speed	15.7	16.2	16.4	15.5	15.0	14.8	13.3	13.8	13.6	13.0
Central + inner area evening-peak speed	14.0	13.9	14.7	13.0	13.5	12.7	11.4	12.2	12.2	11.1

Source: Department of the Environment, Transport, and the Regions, 'Traffic Speeds in Inner London: 1998', *DETR Statistics Bulletin '98*.

One of the problems with these statistics is that the area being surveyed, and the methodology of the survey, changed throughout the period, but at least they help us grasp the order of magnitude of the situation. Thus, assuming the 'central area' was defined more or less the same over the century, that the hours representing the peaks and inter-peak period are also comparable, and taking the values for 1906 from the top average speed of motor-cabs, we can derive a historical pattern represented in Figure 1.1.

Looking at all these tables and figures, it seems that London traffic today really does go no faster than it did in the horse-drawn era. But what has this analysis of the statistics really told us?

The problem with the traffic problem

In 1954, journalist Darrell Huff published *How to Lie with Statistics*, a popular work recently described by a statistician (presumably with a straight face) as 'the most widely read statistics book in the history of the world'.[27] In it, Huff trained his readers to watch out for devices such as 'the well-chosen average' and 'the gee-whiz graph', just two of several techniques employed by people seeking to manipulate statistics to present a particular message (and that might look familiar from the foregoing account of traffic statistics). Huff was not a statistician, and some on the inside of the profession were uncomfortable with the points he made, but most accepted his premise. As one said recently, 'People do lie with statistics every day, and it is to Huff's credit that he takes the media (and others) to task for having stretched, torn or mutilated the truth.'[28]

It is now quite clear that repeatedly changing the sample and the methodology will render any meaningful statistical analysis impossible. While there is no suggestion the GLC, DETR, Road Research Laboratory, or Henry Watson intended to deceive, the fact that they were operating in politically inflected environments, and that they all wanted to convey particular messages with their

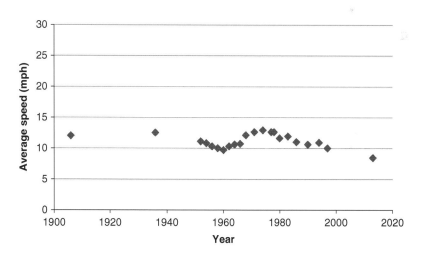

Figure 1.1 Average traffic speed during weekday daytimes for motor traffic in central London, 1906–2013, constructed using sources described in text. Compiled by the author.

statistics, means we should at least consume their fare with doses of salt. The problem is that the traffic problem starts to break down under scrutiny into numerous other problems – human problems, requiring approaches from the humanities as much as the sciences to make sense of them.

Statistics will not tell us the wider story. For this, it is crucial to look at how movement and motion in cities have been conceptualized. The traffic 'problem' is often defined in terms of congestion, but it is really a set of conceptual problems about mobilities, which have geographical, political, and technological aspects. We physically configure the places and spaces of traffic according to values and beliefs. In talking about the traffic problem, we are passing judgement on other people. Ideas underpinning the approach in this book to these three aspects – geographies, politics, and technologies – will now be considered.

Technologies, politics, and geographies

Scholars in the humanities have been turning increasingly towards infrastructure over recent years. In the history of technology, Thomas Hughes established a research programme in 1983 that has since been developed by countless writers seeking a sophisticated understanding of networked technologies. In urban studies and related fields, authors such as Stephen Graham, Maria Kaika, Simon Marvin, Karen Ruhleder, Susan Leigh Star, Erik Swyngedouw, and Nigel Thrift have offered particularly notable contributions. Together, this work has helped situate studies of urban planning, mobility, society, and nature within technological systems, and it is impossible to imagine studies of infrastructure that do not engage critically with material culture. In addition, a recent focus on maintenance and repair, as well as infrastructures in their failed states, whether acute or chronic, shows there is analytical value in considering the fragility of infrastructure. It helps us see that which is normally invisible.[29]

Thus traffic congestion helps us see mobility. From this latter point it would be easy to characterize congestion as the failed state of mobility. Yet the infrastructure turn helps problematize that position. In their study of maintenance, Stephen Graham and Nigel Thrift comment that 'it can be argued that the accidents that stem from so many breakdowns are not aberrant but are a part of the thing itself. To invent the train is to invent the train crash, to invent the plane is to invent the plane crash, and so on.'[30] In this sense, congestion is not a failed state: it is an ineluctable part of mobility, and this reinforces the view that the term 'traffic problem' is itself problematic and deserves to be subjected to critical scrutiny.

This book, then, assesses a material culture of traffic, but defined carefully. Graham and Thrift argue that:

> Cultures of normalized and taken-for-granted infrastructure use sustain widespread assumptions that urban 'infrastructure' is somehow a material and utterly fixed assemblage of hard technologies embedded stably in place,

which is characterized by perfect order, completeness, immanence and internal homogeneity rather than leaky, partial and heterogeneous entities.[31]

They also argue that those infrastructure cultures have histories. Giving the example of how societies deal with car accidents, they comment that 'this crash ecology has not come into existence overnight. It has been built up over a long period of time'.[32] The idea of an 'ecology' points towards networks including things such as skills and knowledge, as well as hard artefacts. In our example, traffic congestion is clearly a juxtaposition of materiality (artefacts, technologies) and people (with skills, knowledges, and beliefs). And it is these ideas about assemblages – actor-networks that comprise and are performed by both human and non-human elements – which can guide us towards a more sophisticated approach to a socio-technical culture of traffic.

Antoine Bousquet has recently summarized the socio-technical nature of assemblages, commenting that:

> the deployment of any technology within the social field must necessarily be grasped in terms of the wider ensemble within which it is inserted, since the multiple relations that compose this ensemble determine to a large extent the usages and meanings given to the said technology.[33]

But what does this general comment mean for a study of traffic and congestion? Bousquet's conclusions give useful summaries of assemblage theory that can directly transfer to 'the traffic problem'. One can note his following comment:

> technology is simultaneously both less and more than what it is typically taken to be: less because it's not an external material agency that unilaterally transforms a passive social body, and more because it actually permeates every aspect of the social. The question of technology thus directs us towards the ubiquitous materiality of social relations, the very glue that holds human collectives together.[34]

If we substitute the word 'congestion' for 'technology' in the quote, it points towards the need to avoid an approach that is pure technological determinism or pure social construction.

Instead we need to consider action and influence in both directions. Work in this field looking specifically at automobility includes that of Peter Merriman, who guides us towards looking at multiple actors and viewpoints. He notes that we need to look at (among other groups) road safety experts, government scientists, police, judiciary, traffic engineers, and commercial organizations, as well as planners, engineers, landscape architects, and preservationists when considering the planning of traffic.[35] He discusses electronic monitoring networks as well as codes of conduct and the practice of conventions. And he has commented explicitly on the significance of technologies in this network:

roads and motorways are not simply products of political decision-making, planning or engineering, for a large number of actors are required to bring about such socio-technical achievements: from human actors such as planners, politicians, engineers, lobbyists, labourers and landscape architects, to distinctive arrangements of concrete, tarmac and steel.[36]

Ideas of traffic and congestion as socio-technical assemblages will therefore run implicitly throughout this book.

So, too, will issues surrounding the peculiarities of political decision-making in the capital. 'Technically there is no London', wrote public affairs scholar Hank Savitch in 1988. He was describing the capital's political governance, divided among Greater London, the City of London, and the London boroughs, and he observed that the system had led to 'the vice of a confused identity'. The result was an 'intractable metropolis'.[37] Michael Hebbert has expanded this analysis, observing that the tension between the London County Council on the one hand and the City Corporation and boroughs on the other hand ('the parochialism of the vestries') was actively created at the turn of the twentieth century by Tory politicians seeking to limit the influence of the Liberal and socialist LCC. This Conservative Corporation and borough opposition to the metropolitan LCC formed, in Hebbert's words, 'an effective and durable political axis' and 'a system of permanent ambiguity in the issue of who should speak and act for London'.[38] As Jerry White has noted, 'rivalry and conflict and controversy were bred in the bones of London governance from the very start of the [twentieth] century'.[39]

The story of London's government has been well and carefully told, and remains a live subject, with the formation of the Greater London Authority in 2000 the latest in a century-and-a-half of attempts to govern 'an ungovernable city', in the words of Tony Travers.[40] Of particular interest in the context of this book is to consider the key driver for unification of London government from the 1850s onwards: infrastructure. Right from the start, Hebbert has argued, 'The great unifier was transport.'[41] The formation of the first capital-wide governance body, the Metropolitan Board of Works (MBW) in 1855, emerged from problems with sewerage and sanitation, but included improvements to the capital's arterial road network in its founding scope, as John Davis has noted.[42] The MBW was, as Travers has described, 'a temporary technical solution to the problems of equipping the world's largest metropolis with basic infrastructure'.[43] Its supersession by the LCC in 1889 continued this work, in part to enact transport improvements so that overcrowding could be eased by an active dispersal policy, as Andrew Saint has argued.[44] Then, as the LCC gave way to the GLC in 1965, 'Arterial roads – or rather the lack of them – were a dominant consideration' for the government, in Hebbert's words. 'The council was created expressly to get the primary road network built', he has noted.[45] Thus transport – particularly road transport – has remained an enduring factor in the ongoing political tension between metropolitan and local government in the capital, and will be a recurring theme throughout the case studies, which will look at how this tension played out, in detail, in the streets themselves.

But of course, the politics inherent in congestion and the traffic problem is wider and deeper than overt governance structures, important though they have been in the shaping of responses to traffic. A wider consideration of the politics of traffic is invited following the work of writers such as David Harvey, Doreen Massey, Patrick Joyce, Gareth Stedman Jones, and Tim Cresswell, who reveal the ways cities and their infrastructures comprise, and constitute, political acts – political in the widest sense of the word, including body politics.[46] These issues, too, will repeatedly emerge in the stories that follow.

Recent writing on mobilities draws together and provides theoretical rigour to these attempts to break down a seemingly fixed, singular, technical 'problem' and decode its politics. This revolves around ideas of flow and fixity, a pairing of concepts that could be tailor-made for treating the subject of traffic congestion. Introducing the concept, Ash Amin and Nigel Thrift encourage a focus on flow rather than stasis, commenting that 'Modern cities are extraordinary agglomerations of flows.'[47] They observe that in their work 'there is a strong emphasis on understanding cities as spatially open and cross-cut by many different kinds of mobilities, from flows of people to commodities and information'.[48] They consider the city as a 'machinic assemblage, which is both composed by, and institutionalizes, flow'.[49] This approach points up a negative value judgement placed on congestion. If Tim Cresswell says movement is mobility without meaning, as Peter Adey has concluded, then we need to scrutinize the *meaning* of congestion, which means seeing it as a form of mobility itself (mobility can incorporate immobility) rather than a meaning-free and uncritical 'lack of movement'.[50] Adey says: 'To abstract mobility to movement (mobility without meaning) often has distinctly political consequences.'[51] So, the abstraction of a wide range of experiences of traffic to a singular traffic problem is the result of a set of political acts.

Melissa Butcher has specifically examined a political materiality of congestion. She notes that 'Physical zones of exclusion such as gated communities and gentrification projects are designed to remove uncivil spaces of congestion that cause discomfort', and that 'Local markets ... are gentrified ... by clean shopping malls that enable the easy flow of traffic within and without. Space becomes privatised, regulated and surveilled to contain or exclude congestion.'[52] However, a risk in this approach is that there appears to be little human agency: spaces 'become' surveilled; gentrification projects 'are designed'; and it is malls that gentrify street markets. A more active view sees people constituting gentrification projects and surveillance systems, and examples of both will be considered in this book. Adey reinforces this point: 'Mobility is not essentially good or bad ... Rather mobility is *given* or inscribed with meaning', depending on context and the people doing the inscribing.[53]

He has also observed that:

> The politics of mobility is rarely a new thing, but often builds on the back of much longer historical trends and of ideologies sedimented into plans and processes, concretized into large-scale building projects and infrastructure. Here these included a general privileging of private automobility as the norm.[54]

Two lessons deserve to be drawn from Adey's comments here. The first is about biography and history. Mobilities have histories and they are constituted by human agency, a point made too by Miles Ogborn in his approach to telling global history when he comments that:

> it is easier to see the effects of human action, and the effects on human action, once that global scale is understood as the on-going organisation of different and partial relationships, networks and webs. These modes of connection may stretch across vast areas, but they always do so via particular forms of action.

The aim of his approach, he observes, 'is to present the world as seen from many points of view'. He notes that his work 'gives due weight to the actions of all sorts of people in exploring these global histories and geographies' and is thus a call for multiple historical biography in the telling of historical geography. As he concludes, 'People can and do make a difference for themselves and others.'[55]

This point can be seen in the context of another 'turn' in the humanities – the biographical turn – which seeks to critically examine the ways in which biographical research and life-writing affect the ways we interrogate and interpret histories. Barbara Caine has commented that the biographical turn 'involves a new preoccupation with individual lives and stories as a way of understanding both contemporary societies and the whole process of social and historical change', and that a biographical approach gives historians 'a way of accessing subjective understanding and experience'.[56] This book will follow that logic, aiming to use individual biographical and historical cases to understand better the subjective experiences of traffic that are occluded by seemingly objective statistical or economic accounts, as well as to gain insights into wider processes of change. In doing so, it attempts to reinforce the value of studying the 'longer historical trends' described by Adey above.

The second point to be made in response to Adey's comments is about scale and geography. By focusing on the large scale of the traffic problem, one is likely to see large-scale infrastructures and large-scale trends, such as the privileging of private automobility, as Adey notes. The significance of both of these in the story that will be told in this book cannot be denied. However, there is value in balancing large-scale historical study with the parallel study of smaller-scale episodes – which might in turn illuminate different large-scale trends. By looking at the fine grain of seemingly localized projects and infrastructures, and demonstrating how the configurations of mobilities have been enacted in specific places at specific times, we can take into account a further point made by Miles Ogborn, with Frank Mort, who together noted the following:

> What is uncovered at close range is much more about the ways in which conceptions of urban society and culture and the geographical ordering of the city are interrelated in the history of specific streets and thoroughfares, monuments, buildings, and even distinctive interiors ... the turn to the detailed particularities of place and setting, and to the multiple social actors who inhabit these environments, produces a particular effect. London's geographies now become active sites for examining the competing uses, social meanings, and power relations that have structured the development of the city.[57]

This book will follow this logic, too, exploring a range of specific London projects in order to interrogate the use of traffic and the streets as sites for the performance of power relations.

Notes

1 Ruth Bashall and Gavin Smith, 'Jam Today: London's Transport in Crisis', in *The Crisis of London*, ed. Andy Thornley (London: Routledge, 1992), 37.
2 *Evening Standard*, 12 November 1938, 3; 14 November, 11; 15 November, 3; 16 November, 14; 17 November, 13; 18 November, 23; 19 November, 19; 21 November, 12; 22 November, 18; 23 November, 12; 24 November, 12; 25 November, 11; 26 November, 15; 28 November, 6 and 15; 29 November, 11.
3 *Evening Standard*, 19 November 1938, 19; 24 November, 12.
4 Charles Bressey and Edwin Lutyens, *Ministry of Transport Highway Development Survey 1937 (Greater London)* (London: HMSO, 1938), 17–18.
5 Bressey and Lutyens, 17.
6 Bressey and Lutyens, 18.
7 *Manchester Guardian*, 18 June 1936, 8. See also, among others, 30 July 1936, 6.
8 Peter Ackroyd, *London: The Biography* (London: Vintage, 2001), 104.
9 Stephen Inwood, *A History of London* (London: Macmillan, 1998), 541, 548.
10 James Winter, *London's Teeming Streets, 1830–1914* (London: Routledge, 1993), 19, 43.
11 Roy Porter, *London: A Social History* (London: Penguin, 2000), 273–4.
12 Jonathan Schneer, *London 1900: The Imperial Metropolis* (New Haven: Yale University Press, 1999), 5–6.
13 Jerry White, *London in the Twentieth Century: A City and Its People* (London: Vintage Books, 2008), 12.
14 Joe Moran, 'Crossing the Road in Britain 1931–1976', *Historical Journal* 49, no. 2 (2006): 478.
15 Simon Gunn, 'People and the Car: The Expansion of Automobility in Urban Britain, *c.* 1955–70', *Social History* 38, no. 2 (2013): 221; Simon Gunn, 'Introduction', in *Traffic in Towns: A Study of the Long Term Problems of Traffic in Urban Areas (The Buchanan Report)*, ed. Colin Buchanan (Abingdon: Routledge, 2015), ix.
16 Henry Watson, *Street Traffic Flow* (London: Chapman and Hall, 1933), v–vii.
17 Watson, 367.
18 T. C. Barker and Michael Robbins, *A History of London Transport*, Vol. II: *The Twentieth Century to 1970* (London: George Allen & Unwin, 1974), Chapters 11, 12, 15, 16.

19 John Halas and Joy Batchelor, dirs., *Automania 2000* (film), 1963; Jez Stewart, 'The Greatest Animated Films of All Time?', *BFI News, Features and Opinion* (blog), 24 April 2014, www.bfi.org.uk/news/greatest-animated-films-all-time.

20 Department for Transport, *An Introduction to the Department for Transport's Road Congestion Statistics* (London: Department for Transport, 2013), italics in original.

21 London County Council, *London Statistics 1936–38* (London: London County Council, 1938), 317.

22 Transport for London, *Travel in London: Report 7* (London: Transport for London, 2014), 145.

23 London County Council, *London Statistics 1905–6* (London: London County Council, 1906), 334. The 'central area' was defined as 'the cities of London and Westminster, the boroughs of St Marylebone, St Pancras (south of Euston-road), Holborn, Finsbury, Shoreditch, Bethnal-green, Stepney, Southwark, Bermondsey, and Lambeth (north of Kennington-lane), an area of 12,268 acres (about 19 square miles)' (24).

24 Greater London Council, *Annual Abstract of Greater London Statistics 1968* (London: Greater London Council, 1968), 61.

25 Greater London Council, *Annual Abstract of Greater London Statistics 1979 and 1980* (London: Greater London Council, 1980), 88. The periods were defined as follows: 7.45 to 9.15 a.m. (morning peak), 10 a.m. to 4 p.m. (off-peak), and 4.45 to 6.15 p.m. (evening peak).

26 Department of the Environment, Transport, and the Regions, 'Traffic Speeds in Inner London: 1998', *DETR Statistics Bulletin '98*, quoted at www.camden.gov.uk/ccm/cms-service/stream/asset/?asset_id=570372.

27 Darrell Huff, *How to Lie with Statistics* (London: Victor Gollancz, 1954); J. Michael Steele, 'Darrell Huff and Fifty Years of *How to Lie with Statistics*', *Statistical Science* 20, no. 3 (2005): 205.

28 Steele, 207.

29 Thomas P. Hughes, *Networks of Power: Electrification in Western Society, 1880–1930* (Baltimore: Johns Hopkins University Press, 1983); Susan Leigh Star and Karen Ruhleder, 'Steps toward an Ecology of Infrastructure: Design and Access for Large Information Spaces', *Information Systems Research* 7, no. 1 (March 1996): 111–34; Susan Leigh Star, 'The Ethnography of Infrastructure', *American Behavioral Scientist* 43, no. 3 (December 1999): 377–91; Maria Kaika and Erik Swyngedouw, 'Fetishizing the Modern City: The Phantasmagoria of Urban Technological Networks', *International Journal of Urban and Regional Research* 24, no. 1 (March 2000): 120–38; Stephen Graham and Simon Marvin, *Splintering Urbanism: Networked Infrastructures, Technological Mobilities and the Urban Condition* (London: Routledge, 2001); Stephen Graham and Nigel Thrift, 'Out of Order: Understanding Repair and Maintenance', *Theory, Culture and Society* 24, no. 3 (2007): 1–25; Stephen Graham, ed., *Disrupted Cities: When Infrastructure Fails* (Abingdon: Routledge, 2010).

30 Graham and Thrift, 4.

31 Graham and Thrift, 10.

32 Graham and Thrift, 17.

33 Antoine Bousquet, 'Welcome to the Machine: Rethinking Technology and Society through Assemblage Theory', in *Reassembling International Theory: Assemblage Thinking and International Relations*, ed. Michele Acuto and Simon Curtis (Basingstoke: Palgrave Macmillan, 2014), 93.

34 Bousquet, 96–7.

35 Peter Merriman, 'Automobility and the Geographies of the Car', *Geography Compass* 3, no. 2 (2009): 590.

36 Peter Merriman, 'Roads', in *The Routledge Handbook of Mobilities*, ed. Peter Adey, David Bissell, Kevin Hannam, Peter Merriman, and Mimi Sheller (Abingdon: Routledge, 2014), 201.

37 Hank Savitch, *Post-Industrial Cities: Politics and Planning in New York, Paris, and London* (Princeton: Princeton University Press, 1988), 170–1.

38 Michael Hebbert, *London: More by Fortune than Design* (Chichester: John Wiley & Sons, 1998), 50–2; see also Michael Hebbert, 'Governing the Capital', in *The Crisis of London*, ed. Andy Thornley (London: Routledge, 1992), 134–48.

39 White, 357.

40 Tony Travers, *The Politics of London: Governing an Ungovernable City* (Basingstoke: Palgrave Macmillan, 2004); see also Tony Travers, *London's Boroughs at 50* (London: Biteback, 2015).

41 Hebbert, *London*, 39.

42 John Davis, *Reforming London: The London Government Problem 1855–1900* (Oxford: Oxford University Press, 1988), 41; see also Gloria Clifton, *Professionalism, Patronage and Public Service in Victorian London: The Staff of the Metropolitan Board of Works 1856–1889* (London: Athlone Press, 1992).

43 Travers, *The Politics of London*, 25.

44 Andrew Saint, '"Spread the People": The LCC's Dispersal Policy, 1889–1965', in *Politics and the People of London: The London County Council 1889–1965*, ed. Andrew Saint (London: Hambledon Press, 1989), 215–36.

45 Hebbert, *London*, 76.

46 See for instance David Harvey, *Social Justice and the City* (London: Edward Arnold, 1973); David Harvey, *Paris, Capital of Modernity* (London: Routledge, 2003); Doreen Massey, *Space, Place, and Gender* (Minneapolis: University of Minnesota Press, 1994); Doreen Massey, *For Space* (London: Sage, 2005); Patrick Joyce, *The Rule of Freedom: Liberalism and the Modern City* (London: Verso, 2003); Patrick Joyce, *The State of Freedom: A Social History of the British State since 1800* (Cambridge: Cambridge University Press, 2013); Gareth Stedman Jones, *Outcast London: A Study in the Relationship between Classes in Victorian Society* (Oxford: Oxford University Press, 1971); Tim Cresswell, *On the Move* (New York: Routledge, 2006); Tim Cresswell, *Place: An Introduction* (Chichester: John Wiley & Sons, 2015).

47 Ash Amin and Nigel Thrift, *Cities: Reimagining the Urban* (Cambridge: Polity Press, 2002), 42.

48 Amin and Thrift, 3.

49 Amin and Thrift, 5.

50 Peter Adey, *Mobility* (Abingdon: Routledge, 2010), 34–5.

51 Adey, *Mobility*, 35.

52 Melissa Butcher, 'Congestion', in Adey *et al.*, *The Routledge Handbook of Mobilities* (Abingdon: Routledge, 2014), 462.

53 Adey, *Mobility*, 36.

54 Peter Adey, 'Mobilities: Politics, Practices, Places', in *Introducing Human Geographies*, 3rd edn, ed. Paul Cloke, Philip Crang, and Mark Goodwin (Abingdon: Routledge, 2014), 796.

55 Miles Ogborn, *Global Lives: Britain and the World, 1550–1800* (Cambridge: Cambridge University Press, 2008), 8–10.

56 Barbara Caine, *Biography and History* (New York: Palgrave Macmillan, 2010), 1; and see also the essays in Hans Renders, Binne de Haan, and Jonne Harmsma, eds., *The Biographical Turn: Lives in History* (Abingdon: Routledge, 2017), which offer a comprehensive survey of work in the field.

57 Frank Mort and Miles Ogborn, 'Transforming Metropolitan London, 1750–1960', *Journal of British Studies* 43, no. 1 (January 2004): 4–5.

Bibliography

Evening Standard.

Manchester Guardian.

Ackroyd, Peter. *London: The Biography.* London: Vintage, 2001.

Adey, Peter. 'Mobilities: Politics, Practices, Places'. In *Introducing Human Geographies.* 3rd edn, ed. Paul Cloke, Philip Crang, and Mark Goodwin, 791–805. Abingdon: Routledge, 2014.

———. *Mobility.* Abingdon: Routledge, 2010.

Amin, Ash, and Nigel Thrift. *Cities: Reimagining the Urban.* Cambridge: Polity Press, 2002.

Barker, T. C., and Michael Robbins. *A History of London Transport.* Vol. II: *The Twentieth Century to 1970.* London: George Allen & Unwin, 1974.

Bashall, Ruth, and Gavin Smith. 'Jam Today: London's Transport in Crisis'. In *The Crisis of London*, ed. Andy Thornley, 37–55. London: Routledge, 1992.

Bousquet, Antoine. 'Welcome to the Machine: Rethinking Technology and Society through Assemblage Theory'. In *Reassembling International Theory: Assemblage Thinking and International Relations*, ed. Michele Acuto and Simon Curtis, 91–141. Basingstoke: Palgrave Macmillan, 2014.

Bressey, Charles, and Edwin Lutyens. *Ministry of Transport Highway Development Survey 1937 (Greater London).* London: HMSO, 1938.

Butcher, Melissa. 'Congestion'. In *The Routledge Handbook of Mobilities*, ed. Peter Adey, David Bissell, Kevin Hannam, Peter Merriman, and Mimi Sheller, 460–7. Abingdon: Routledge, 2014.

Caine, Barbara. *Biography and History.* New York: Palgrave Macmillan, 2010.

Clifton, Gloria. *Professionalism, Patronage and Public Service in Victorian London: The Staff of the Metropolitan Board of Works 1856–1889.* London: Athlone Press, 1992.

Cresswell, Tim. *On the Move.* New York: Routledge, 2006.

———. *Place: An Introduction.* Chichester: John Wiley & Sons, 2015.

Davis, John. *Reforming London: The London Government Problem 1855–1900.* Oxford: Oxford University Press, 1988.

Department for Transport. *An Introduction to the Department for Transport's Road Congestion Statistics.* London: Department for Transport, 2013.

Department of the Environment, Transport, and the Regions, 'Traffic Speeds in Inner London: 1998'. In *DETR Statistics Bulletin '98*, quoted at www.camden.gov.uk/ccm/cms-service/stream/asset/?asset_id=570372.

Graham, Stephen, ed. *Disrupted Cities: When Infrastructure Fails.* Abingdon: Routledge, 2010.

Graham, Stephen, and Simon Marvin. *Splintering Urbanism: Networked Infrastructures, Technological Mobilities and the Urban Condition.* London: Routledge, 2001.

Graham, Stephen, and Nigel Thrift. 'Out of Order: Understanding Repair and Maintenance'. *Theory, Culture and Society* 24, no. 3 (2007): 1–25.

Greater London Council. *Annual Abstract of Greater London Statistics 1968.* London: Greater London Council, 1968.

———. *Annual Abstract of Greater London Statistics 1979 and 1980.* London: Greater London Council, 1980.

Gunn, Simon. 'Introduction'. In *Traffic in Towns: A Study of the Long Term Problems of Traffic in Urban Areas (The Buchanan Report)*, ed. Colin Buchanan, viii–xvi. Abingdon: Routledge, 2015.

———. 'People and the Car: The Expansion of Automobility in Urban Britain, *c.* 1955–70'. *Social History* 38, no. 2 (2013): 220–37.

Halas, John, and Joy Batchelor, dirs. *Automania 2000* (film), 1963.

Harvey, David. *Paris, Capital of Modernity.* London: Routledge, 2003.

———. *Social Justice and the City.* London: Edward Arnold, 1973.

Hebbert, Michael. 'Governing the Capital'. In *The Crisis of London*, ed. Andy Thornley, 134–48. London: Routledge, 1992.

———. *London: More by Fortune than Design.* Chichester: John Wiley & Sons, 1998.

Huff, Darrell. *How to Lie with Statistics.* London: Victor Gollancz, 1954.

Hughes, Thomas P. *Networks of Power: Electrification in Western Society, 1880–1930.* Baltimore: Johns Hopkins University Press, 1983.

Inwood, Stephen. *A History of London.* London: Macmillan, 1998.

Joyce, Patrick. *The Rule of Freedom: Liberalism and the Modern City.* London: Verso, 2003.

———. *The State of Freedom: A Social History of the British State since 1800.* Cambridge: Cambridge University Press, 2013.

Kaika, Maria, and Erik Swyngedouw. 'Fetishizing the Modern City: The Phantasmagoria of Urban Technological Networks'. *International Journal of Urban and Regional Research* 24, no. 1 (March 2000): 120–38.

London County Council. *London Statistics 1905–6.* London: London County Council, 1906.

———. *London Statistics 1936–38.* London: London County Council, 1938.

Massey, Doreen. *For Space.* London: Sage, 2005.

———. *Space, Place, and Gender.* Minneapolis: University of Minnesota Press, 1994.

Merriman, Peter. 'Automobility and the Geographies of the Car'. *Geography Compass* 3, no. 2 (2009): 586–99.

———. 'Roads'. In *The Routledge Handbook of Mobilities*, ed. Peter Adey, David Bissell, Kevin Hannam, Peter Merriman, and Mimi Sheller, 196–204. Abingdon: Routledge, 2014.

Moran, Joe. 'Crossing the Road in Britain 1931–1976'. *Historical Journal* 49, no. 2 (2006): 477–96.

Mort, Frank, and Miles Ogborn. 'Transforming Metropolitan London, 1750–1960'. *Journal of British Studies* 43, no. 1 (January 2004): 1–14.

Ogborn, Miles. *Global Lives: Britain and the World, 1550–1800.* Cambridge: Cambridge University Press, 2008.

Porter, Roy. *London: A Social History.* London: Penguin, 2000.

Renders, Hans, Binne de Haan, and Jonne Harmsma, eds. *The Biographical Turn: Lives in History.* Abingdon: Routledge, 2017.

Saint, Andrew. '"Spread the People": The LCC's Dispersal Policy, 1889–1965'. In *Politics and the People of London: The London County Council 1889–1965*, ed. Andrew Saint, 215–36. London: Hambledon Press, 1989.

Savitch, Hank. *Post-Industrial Cities: Politics and Planning in New York, Paris, and London.* Princeton: Princeton University Press, 1988.

Schneer, Jonathan. *London 1900: The Imperial Metropolis.* New Haven: Yale University Press, 1999.

Star, Susan Leigh. 'The Ethnography of Infrastructure'. *American Behavioral Scientist* 43, no. 3 (December 1999): 377–91.

Star, Susan Leigh, and Karen Ruhleder. 'Steps toward an Ecology of Infrastructure: Design and Access for Large Information Spaces'. *Information Systems Research* 7, no. 1 (March 1996): 111–34.

Stedman Jones, Gareth. *Outcast London: A Study in the Relationship between Classes in Victorian Society*. Oxford: Oxford University Press, 1971.

Steele, J. Michael. 'Darrell Huff and Fifty Years of *How to Lie with Statistics*'. *Statistical Science* 20, no. 3 (2005): 205–9.

Stewart, Jez. 'The Greatest Animated Films of All Time?'. *BFI News, Features and Opinion* (blog), 24 April 2014. www.bfi.org.uk/news/greatest-animated-films-all-time.

Transport for London. *Travel in London: Report 7*. London: Transport for London, 2014.

Travers, Tony. *London's Boroughs at 50*. London: Biteback, 2015.

———. *The Politics of London: Governing an Ungovernable City*. Basingstoke: Palgrave Macmillan, 2004.

Watson, Henry. *Street Traffic Flow*. London: Chapman and Hall, 1933.

White, Jerry. *London in the Twentieth Century: A City and Its People*. London: Vintage, 2008.

Winter, James. *London's Teeming Streets, 1830–1914*. London: Routledge, 1993.

2 The traffic problem in urban planning

Introduction

The introduction and rise of the motor vehicle and of professional urban planning share a common chronology. The Locomotives on Highways Act of 1896 unlocked the potential of motor vehicles as a form of mass transport; two years later Ebenezer Howard published his ideas for Garden Cities, and by the mid-1900s town planning had been established as a discipline. It makes sense, then, that the eyes of historians and geographers have most commonly turned towards the planning discipline to find characterizations of, and understand responses towards, traffic congestion in the motor age. This chapter thus examines the traffic problem as seen in the professional urban planning discipline. It surveys a series of canonical London plans from the 1910s to the 1970s, considering how each successive plan and its planners characterized traffic and proposed to deal with it. This is done in order to appreciate the dominant framing of traffic as a problem with a particular and canonical set of solutions.

But by treating traffic as an assemblage or socio-technical network, this reading will also help identify figures, forces, projects, and world views that existed in the margins of mainstream urban planning, either actively marginalized owing to their incompatibility with urban planning approaches, or acting as influences on urban planning from outside the formal discipline. One such actor, for instance, is capital, which has had a particularly complex relationship with traffic and planning. Some of these marginal approaches, having been identified in this chapter's conclusion, will be brought into central view in later chapters, offering as a whole a wider and more complex account of traffic congestion in twentieth-century London.

The literature on the history of urban planning is, of course, extensive. London, as capital city, has very often been the focus for attention, and the years before and after the Second World War have commonly been seen as catalysts for the planned solution of the traffic problem, particularly given the scale of destruction of London from bombing. Hermione Hobhouse, for instance, put 'traffic flow or rather its congestion' first in a list of 'historic problems' that wartime planning sought to solve, noting it as 'One of the most all-pervading concerns' throughout the history of London planning.[1] Gordon Cherry observed a growing interest in tackling transport congestion through planning in the 1930s, after decades of piecemeal localized road improvements, although it was not until the Second World War that the interest intensified, and the 1960s that demand really took off.[2] Donald Foley had situated this concern in decentralization proposals, which he showed were seen as means to reduce the amount of daily commuting.[3] Helen Meller described 'the revolution in transport' in the early 1960s, 'which seemed to have had most implications for the physical environment of cities'.[4] For Douglas Hart, voicing early 1960s concerns, traffic congestion was causing planners to 'lose control of the capital'.[5]

Given this general acceptance of the crucial significance of traffic congestion to the development of the twentieth-century planning of London, it seems that a good starting point in studying London congestion per se is to examine plans for London and chart their characterizations of the problem. However, this raises interesting issues. The dominant historiography of London planning focuses on a limited set of what might be termed canonical plans and their (named) authors. But the question of what constitutes a plan, who is a planner, what authorship of a plan means in practice, and how one might build a hierarchical relationship as well as a chronological or genealogical one between them all, is by no means trivial.

These issues have been considered at length by Peter Larkham and David Adams, whose work has particularly informed this chapter. They note the general focus in histories of urban reconstruction on: '"great plans" and "great planners" to the virtual exclusion of all other forms of reconstruction planning: one might think that Abercrombie alone was responsible for post-war reconstruction planning!'[6] They also note the porous disciplinary boundaries defining the term 'planner' and observe that architects, engineers, and surveyors often carried out planning work. They both invite and offer a wider view, and the plans chosen for the following survey, at least those from the particularly intensive period of 1940–7, draw largely on Larkham and Adams's own selection.

By looking for the politics, technologies, and geographies embedded in the authors' visualizations of congestion and its solution, one sees that circulation and traffic have been at the heart of professional town planning throughout the twentieth century, and a set of common themes and actors are encountered. These themes include geographies of governance and the relationship between London-wide bodies and local authorities; technological visions of grids, networks, verticality, and progress; social geographies of segregation and control; and the politics and financing of roads and mobility. Common actors include the big names of Unwin, Abercrombie, Holford, and Buchanan but also (responding to Larkham and Adams's invitation) figures such as the police traffic commissioner Alker Tripp and the engineer Charles Bressey, who are less commonly associated with London's mid-century planning landscape.

The birth of town planning, 1898–1910

London's traffic problem was not a modern one. Its streets were teeming long before the motor-car, as James Winter has demonstrated.[7] What is of interest here is the modernization of the response to the city and its problems. At the turn of the twentieth century, several interlinked factors – decentralization, motorization, and the growth of government intervention in social welfare and the physical environment – came together to create a new set of responses that focused on planning for a better future, as Helen Meller has described.[8] This pioneering movement fused technological intervention (in the widest sense) with social and environmental reform, and found early expression in Ebenezer Howard's pioneering *Tomorrow*, published in 1898.[9]

Tomorrow set out the now-familiar blueprint of decentralization via fast transport links to 'Garden Cities', themselves models of techno-utopias designed in concentric ring formation for circulation of traffic. In 1899 Howard founded the Garden City Association, which eventually became the Town and Country Planning Association. In 1909, government legislation put town planning on a formal basis and a university course in the subject was offered. In 1910 the first town planning inspector – Thomas Adams – started work, and a journal, the *Town Planning Review*, was launched. Four years later, the Town Planning Institute – a professional body for the new profession of town planner – was formed.[10]

Town planning was an international movement, although with distinctive contributions from practitioners in different countries. As Howard was promoting Garden Cities back home, in Chicago Daniel Burnham was urging the 'City Beautiful', a movement for the creation of grandeur and beauty in American cities such as Chicago and Washington, DC to instil moral, civic virtues in citizens.[11] This was a neo-baroque grandeur of wide boulevards leading to monumental civic centres that, as Jane Jacobs described, performed a 'sorting out' of cultural functions, 'decontaminating their relationship with the workaday city' in a way that harmonized well with Howard's Garden Cities.[12]

Both Garden Cities and Beautiful Cities were thus segregated cities, and a third modern city vision soon emerged: Le Corbusier's *Ville radieuse*, or 'Radiant City', devised in the 1920s.[13] This was a city of high-rise tower blocks separated by expanses of parkland; or rather, the city *was* a park, horizontally, from which the vertical city emerged. Again, fast transport links provided connection and segregation. Jane Jacobs has situated Le Corbusier's Radiant City as a direct product of the Garden City, with traffic at the heart of both visions, as follows:

> He attempted to make planning for the automobile an integral part of his scheme, and this was, in the 1920s and 1930s, a new, exciting idea. He included great arterial roads for express one-way traffic. He cut the number of streets because 'crossroads are an enemy to traffic'. He proposed underground streets for heavy vehicles and deliveries, and of course like the

Garden City planners he kept pedestrians off the streets and in the parks. His city was like a wonderful mechanical toy.[14]

The notion of the city as a giant park had seen earlier expression in the parkways movement of the landscape architect Frederick Law Olmsted, who had conceived of scenic urban roads for pleasure in the horse-drawn era. After motorization, the movement developed into the construction of limited-access motor roads in scenic or landscaped settings, and found ultimate expression on the New York parkways of Robert Moses in the 1920s and beyond.[15]

Carmen Hass-Klau has considered the history of pedestrianism in depth, and concurs with this idea of an international movement advocating pedestrian separation from vehicles, achieved through town planning, to reduce traffic congestion as well as improve safety.[16] Thus, this rapid survey of early town planning clearly shows the importance of roads, and of traffic circulation, flow, and segregation within modern visions of urban existence, and there are clear connections between the approaches of different actors in different places and times. Back in London, the modernizing city was projecting its future as the town planning discipline moved on from its pioneering phase.

The London Society plans, 1918–1921

The London Society was founded in 1912 as a civic amenity society lobbying for a beautiful, ordered city, with a strong focus on traffic and circulation. As Helena Beaufoy has argued, it became an important body in the development of modern urban planning in the interwar period, playing a major role in developing holistic plans for an arterial road network for the capital that directly influenced a young Patrick Abercrombie.[17] Lucy Hewitt has further tracked the idea of the society's involvement in promoting an arterial- and ring-roads network, originating in a 1910 proposal by architect David Barclay Niven.[18] She describes the seminal Royal Institute of British Architects (RIBA) conference on town planning held in London later that year, at which Niven's ring road was proposed by Garden-City doyen George Pepler, and notes that 'It was ... the challenge of providing for London's increasing traffic flow that raised the most pressing concern at the conference.'[19] Out of the meeting emerged the idea of a membership society – which became the London Society – that could lobby for London-wide action. Niven was a founding member, and in 1914 the society joined with RIBA in pressing for the establishment of a single body to design a main road network for the capital.[20] The outcome of this and other government lobbying was the establishment of an arterial roads conference programme held from 1915 onwards.[21]

The London Society's role was substantial, in no small part owing to its publicity activities, and it acted, in David Gilbert's words, as 'a kind of early twentieth-century urban think tank'.[22] Throughout the First World War, the society developed its first London-wide plan, entitled *A Development Plan for Greater London*: a large-scale map depicting the proposed arterial road network and the society's own ideas for parks, parkways, and other open spaces.[23] The map was first revealed in 1918 by the society's founding chairman, architect Aston Webb, before being published by Stanford, exhibited at King's College in April 1919, and later going on permanent public display in the London Museum at Lancaster House, St James's.[24] Figure 2.1 shows its key map.[25] Traffic congestion was the problem front and centre in the society's minds as they created it. In his presentation, Webb stated that 'the main roads round the Metropolis, with their rapid, heavy traffic, are proving totally inadequate'. By 1914, he noted, the capital was still 'without an authorized comprehensive scheme for road improvement'. He ended by offering the map as the first plan 'showing a complete scheme for the improvement and development of Greater London'.[26]

As Hewitt observed, the *Development Plan* demonstrated the society's key argument: that London's traffic infrastructure was critical, and that it needed treatment as a whole. For Gilbert, it appeared at first disappointingly conservative, omitting the central area completely and acting more as a map of record than proposing a new plan for the city, as Patrick Abercrombie's 1940s plans were later to do. But Gilbert draws attention to the map's role in reconfiguring

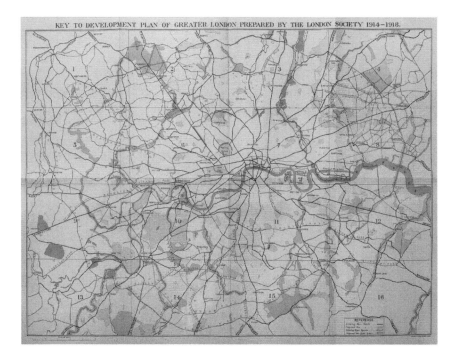

Figure 2.1 Key map to the London Society's *Development Plan of Greater London*, 1918, showing proposed network of new main roads (courtesy of the London Society).

London in the public and professional debate 'as a metropolitan region, rather than as a compact city', being influential in discussions about decentralization and the high-speed motor roads that would make it possible.[27] It was influential, too, in debates about London's government: in 1922, a Royal Commission looking into the subject drew directly from its proposals.[28]

The arterial roads programme, which the society had so strongly supported, began to be built in 1919 as an unemployment-relief project, with some 300 miles constructed by the mid-1930s.[29] The society's mapping work continued through the 1920s with a series of plans of central London showing existing land use, although this did not result in a map or model showing proposed future land usage or road layouts, work that was taken over by the London County Council from 1925.[30]

Yet the society was not just a group of data-gatherers; it had an imaginative, speculative side too. In 1921, the society published its second major planning statement, *London of the Future*, a collection of essays edited by Aston Webb.[31] Described by Gilbert as 'not so much a coherent master plan as an exercise in the contemporary urban imagination' and 'an attempt to rethink London as a city of the twentieth century', the report nevertheless demonstrated the society's overriding concern about traffic and circulation.[32] Five of its 18 essays were overtly concerned with London's transport infrastructure, and all the others focused on it to a greater or lesser extent, characterizing traffic congestion as a threat to ideals of beauty and dignity. This was, too, a technological characterization of congestion. As Gilbert has noted, the book expressed both 'technological enthusiasms' such as those in chapters on aviation and the Channel Tunnel, and 'planning itself as a progressive technology with the power to transform urban existence' in the work of Raymond Unwin and Stanley Adshead.[33]

The London Society showed a rising generation of professional planners the value in both detailed statistical and graphic presentation and in imaginative visions of future Londons. For Patrick Abercrombie in 1920, then a professor at Liverpool University, the success of the *Development Plan* showed what city-wide amenity bodies could achieve, and he never lost touch with the work of the society throughout the 1920s and 1930s, as Hewitt has observed. By the time he was approached in 1941 to work on what became the *County of London Plan* (1943) and *Greater London Plan* (1944) he was explicit in his debt to the society's activities and methodological approach.[34]

Greater London Regional Planning Committee, 1929–1933

One critical figure in the London Society's planning activity was the architect and planner Raymond Unwin, who had made his name in the Garden Cities and suburbs movement. He had organized the 1910 RIBA town planning conference that had led to the society's establishment, and both part-financed and co-directed (with Aston Webb) the *Development Plan of Greater London* project.[35] His essay in the 1921 collection was, as David Gilbert notes, a means to introduce professional planning practice – heavily influenced by work in the United States – to a lay British audience, and promoted a vision of a planned, expanded London based on fast motor roads connecting the decentralized population to employment at the centre.[36] During the 1920s he worked on housing policy with health ministers including Neville Chamberlain. Then, after Chamberlain founded the Greater London Regional Planning Committee (GLRPC) in 1927, Unwin became its technical adviser in 1929 and wrote its landmark *Reports*, the first published the same year and the second four years later.[37]

At the GLRPC, Unwin's approach to traffic congestion continued to be framed in the London Society's discourse of beauty through order and a holistic approach to planning, with an assumption that decentralization of residence – made possible by new transport technologies – would decongest the central area in time. But his focus had by this time turned from the creation of main and arterial roads – now mostly built – to the buildings alongside them. The rapid development of dwellings, shops, and industrial facilities along the new roads – ribbon development – was causing both congestion and a loss of amenity. Congestion, he argued, was caused by obstructions such as frequent junctions and parking outside shopping parades as much as by road width, and evidence from the USA, which was further along the road in this field, was that main roads could either be traffic highways or built-up areas – not both. This was the year after Clough Williams-Ellis had laid into the British approach to ribbon development and eulogized the parkways of the USA.[38]

For Unwin, there were three consequences arising from ribbon development. Firstly, it caused economic inefficiency, as the individual freedoms of frontagers carried a general cost in congestion. This economic argument will be explored in a later chapter. Unwin's second argument was the danger to life and limb owing to accidents arising from fast cars, numerous junctions, and 'the crossings by thoughtless children [and] short-sighted or hesitating folk'. The solution was a form of segregation (although Unwin did not use that word) of fast motor traffic from 'the obstructions and the dangers incidental to frontage development', and these ideas too will be examined in a later chapter. The third consequence of ribbon development, and the one perhaps most distressing to Unwin and the other London Society aesthetes, was that it 'injure[d] the Amenities', bringing town aesthetics into the countryside.[39]

By the time of the second report in 1933, Britain was in the grip of the Great Depression, which was just beginning as the first report was published. Hardest hit were the heavy industrial areas of the Midlands and the north, whereas the new light industries clustering in London and the south-east had already helped to catalyse the development of the main road network. Whilst the depression briefly slowed the rise of car ownership in the early 1930s, a steady rise in the fortunes of the motor industry both drew from and fed a marked expansion of motor traffic in the capital as the decade progressed. Combined with other factors such as migration and large-scale electrification, London's population was booming and the ribbon development along the main roads showed no signs of abating.

Unwin lamented that many of the new main roads were 'now unsuitable for the demands already made on them', with some 'rapidly becoming as congested as some of the areas the road was constructed to by-pass'.[40] Congestion was an indicator of bad or piecemeal planning control, and the solutions almost inevitably came in the form of physical separation of motor-vehicles from pedestrians combined with standardized traffic control and engineering conventions, and new technological forms of traffic junction offering one-way gyratory flow or grade separation, all borrowed from US experience.[41]

Unwin then put forward a list of 71 road improvements needed in Greater London – some already on the planners' radar, others newly proposed – all designed either to create connections between congested areas (many industrial, including the docks) and the main arterial roads, or to bypass existing congestion black-spots or regional centres.[42] But, as with the arterial roads programme itself, none of the proposed schemes entered the central, built-up area – and, unlike the arterial roads and their unemployment-relief financing, no government funds were forthcoming to build Unwin's schemes. The roads campaigner Rees Jeffreys, an insider in the debates of the time, explained why. 'The national financial crisis was assigned by the Government as a reason for deferring any approved works … Once again the Government, in default of action, fell back upon an Inquiry.'[43] It was an engineer, not a planner, who was given the task of preparing a plan for London's roads.

Highway Development Survey, 1937

In December 1934, the Minister of Transport, Leslie Hore-Belisha, instructed Charles Bressey, his Chief Engineer, to prepare a comprehensive survey of the highway developments needed in London over the subsequent 30 years. For the next three years, Bressey worked alongside the architect Edwin Lutyens in a systematic analysis of London's traffic problem.[44] Lutyens had been one of the founder members of the London Society back in 1912 but Bressey was not a planner in any disciplinary sense. Yet this was the first systematic study of London's traffic problem since the dawn of the motor age and the Royal Commission on London Traffic's reports of 1903–5.[45] In fact, at least for the County of London, Bressey observed a total disconnection between official town planning and traffic, stating that:

> as far as the London County Council area is concerned, town planning operations may exercise little direct influence on highway development seeing that the County Council have so far pursued the policy of excluding from their plan all new roads and road improvements.[46]

Bressey's plan was the first to examine systematically what he termed 'centres of congestion' in the Inner London area, namely the 14 worst junctions for causing congestion across the city, acknowledging that new roads were only part of the answer to London's traffic problem.[47] It offered exhaustive analysis of junctions with new main and arterial roads, proposing vertical separation at some, and it examined the growing problem of parked cars. It incorporated large-scale maps pinpointing some 66 proposed road schemes in both Greater London and Inner London – a complex tangle of lines showing the comprehensiveness of Bressey's planning (the central portion of the Greater London plan is shown in Figure 2.2).[48] He even included plans for the London end of a new national motorway network, which was often proposed by those who had experienced the Italian *autostrade* and the German *Reichsautobahnen*.[49] And, in another first for traffic planning, the *Survey* was based on the results of real surveys, as was noted in Chapter 1: observations throughout summer 1936 of traffic delays on key routes, which showed that average speeds in the capital varied from 5.85 mph in the inner zone to 18 mph in the outskirts.[50]

Throughout, the report presented the world view of a civil engineer. Congestion would be solved, and efficient flow would be restored, by building, bridging,

Figure 2.2 Proposed road schemes in Greater London (detail), *Highway Development Survey*, 1937 (Ministry of Transport).

smoothing, connecting, and widening. Town planning principles were left to the local authorities to deal with.[51] Lutyens's role seems likely to have been confined to the architectural treatment of the new flyovers and roundabouts.[52] But, in its further desire to solve congestion by controlling, segregating, and directing the traffic, it also represented a Metropolitan Police world view, for Bressey acknowledged its extensive assistance in his work. He did not name his source there, but it was certainly Alker Tripp, who had been promoted to Assistant Commissioner with responsibility for traffic in 1932, and who, between then and his retirement in 1947, studied and wrote widely on the relationship between road traffic and town planning.

Tripp's role as a transport planner has long been observed.[53] But he was more than an isolated figure at the margins of the discipline: some scholars have noted the influence of his ideas on urban planning more widely. He promoted numerous ideas on the reduction of traffic congestion that were realized in postwar plans for London. These included urban motorways as well as the segregation of pedestrians from motor traffic, either vertically, through raised concrete walkways, or horizontally, through his notion of the 'precinct'. These ideas were encapsulated in his 1942 book *Town Planning and Road Traffic*, for which the foreword was written by Patrick Abercrombie, who said he had 'benefited enormously' from Tripp's work, a fact noted later by Peter Hall and Stephen Ward among others.[54] This joined Tripp's 1938 work *Road Traffic and Its Control* in influencing both European and US planning circles.[55] Several authors have also noted the debt owed to Tripp by Colin Buchanan in his 1963 work *Traffic in Towns*, which will be examined presently.[56] So, whilst neither Bressey's *Highway Development Survey* nor Tripp's *Town Planning and Road Traffic* emerged from within the formal town planning discipline, both underpinned the postwar plans of Abercrombie and Buchanan.

MARS Group, RIBA, and the Royal Academy, 1942–1944

The outbreak of the Second World War intervened in Bressey's plans for a comprehensive road-building programme for London, but did not prevent planning work from continuing. As is well known, after the Blitz attacks on the capital in 1940 and 1941, many planners, ministers, and other interested parties saw opportunities for comprehensive reconstruction along modern planned lines. Before considering the plans of Abercrombie, however, it will be instructive to consider congestion as characterized in a series of 'independent' wartime schemes for London. Emmanuel Marmaras and Anthony Sutcliffe have located the period from 1941 to 1947 as crucial for the development of urban planning policy, the latter date being the introduction of statutory town planning under the 1947 Act.[57] They explore three plans published by independent bodies in 1942 and 1943 that, they suggest, have been overshadowed by the official *County of London* and *Greater London* plans of 1943 and 1944. These plans were published by the Modern Architectural Research (MARS) Group, the Royal Academy (RA), and the RIBA.

The MARS Group was founded in 1933 to promote modern architecture in Britain. It began working on a London urban plan in 1938, rekindling its interest in 1941 as the government requested reconstruction plans following the Blitz bombings. Its work was led by architect and planner Arthur Korn, whose early influences included Raymond Unwin, and the structural engineer Felix Samuely, who took a particular interest in transport and economics.[58] Their proposals were presented at an exhibition in 1942, followed by a written account in which they insisted that 'London must be considered in its entirety, as a working organism.'[59]

The MARS plan was, in Marmaras and Sutcliffe's words, 'clearly not practical, [but] presented an ideal vision of London as it might be if a completely fresh start could be made using all the potential of the "Machine Age"'. And it was a London that Korn and Samuely saw 'almost entirely in terms of movement and distribution', with high-speed transport arteries dominating a total linear rebuilding of the city's boroughs onto a grid to reduce distribution costs.[60] (In this gridded linearity, Marmaras and Sutcliffe note Korn and Samuely's debt to Le Corbusier's *Ville radieuse*, which will figure in a later chapter.[61]) Korn and Samuely recognized that 'the idea of such a superimposed grid may appear somewhat Utopian', but felt radical action was needed as 'any purely local planning is doomed to failure'.[62] In fact, as John Gold has observed, the authors felt that 'the legacy of the past left London with a congested, outmoded and inefficient transport system', which led to their foregrounding plans for circulation and transport.[63]

The transport network in the MARS Group's new London was a model of high-tech segregation: industrial areas segregated from residential and administrative; goods in transit kept out of built-up areas; parcel deliveries located underground; ribbon development along traffic arteries eradicated. All this segregation was required for the following reasons:

> the present-day street system, which is not more advanced than that of old Babylon, cannot cope with twentieth century traffic. This street system causes those obstacles which prevent steady flow, e.g. level crossings and forks, and the presence on the same road of vehicles of all speeds, the slowest of which, naturally, sets the pace.

Traffic lights were out of favour, as 'they have slowed up the traffic even further'; vertical separation at junctions was a better solution.[64] Two-level streets separating goods- and passenger-traffic on shopping streets were an option. Escalators carried pedestrians between levels. Viaducts and one-way systems eradicated conflict. London became very much a machine for circulation.

Idealistic though it certainly was (and was intended to be), this provocative vision of a Modern London, based on an efficient transport grid, was influential in setting a strong modern direction for much wartime and postwar planning of London, although Gold has warned us against considering it as a single plan with only two authors, rather seeing it as expressing the 'experimental and pluralistic nature' of modernist planning thought in this period, which evolved over several years with a multiplicity of voices and approaches.[65]

The second of Marmaras and Sutcliffe's independent plans was a pair of reports from the RA's planning committee.[66] Yet they cast the notion of independence into relief, for whilst the RA was clearly not a statutory body, the membership of its committee showed it to be intimately connected to official planning debates. The RA plans emerged directly from Bressey and Lutyens's *Highway Development Survey* for the Ministry of Transport. As Marmaras and Sutcliffe suggest, the *Survey* had little in it of Lutyens's influence, and in 1940 the RA sought to redress the balance by inviting Lutyens to chair a committee to develop the architectural side of the *Survey*.[67] But it was not merely a committee of architects or Academicians. It included Charles Bressey, Patrick Abercrombie, and Alker Tripp and, as such, needs to be examined more closely than as 'a period piece of academic nostalgia incorporating every cliché in the *Beaux Arts* repertoire', as Lionel Esher described it, or as having value 'more in publicizing an alternative to the MARS Plan than in the significance of its recommendations', as Marmaras and Sutcliffe put it.[68]

In fact, underneath the overt aesthetic nostalgia of the RA plans can be found some powerfully modern statements on transport and circulation that echo some MARS ideas and show the clear controlling hand of Tripp. The second report began by stating that 'The road problem has never been tackled seriously in this country ... That a radical operation is necessary to save the patient from paralysis and strangulation is obvious to all who have given any thought to the subject.'[69] Key to the RA's proposals for traffic were the following: firstly, a high-speed ring road around Inner London dedicated to motor traffic, sunk in a wide cutting clear of existing roads and junctions and connected to the street network by ramps; secondly, a system of vertically segregated arterial roads through the centre and into the suburbs; thirdly, vertically segregated roundabouts incorporating shopping- and business-centres, carrying arterial roads underground and separating pedestrians from vehicles; fourthly, a network of parking spaces and underground garages; and fifthly, a City ring road.

By comparing these traffic proposals with the ring-and-artery network and systematic vertical segregation of the MARS scheme, it is clear that the RA's vision for solving the traffic problem – driven by Tripp and his acolytes, Bressey and Abercrombie – was a transportation machine just as modern as that of Korn and Samuely. The report claimed that 'The adoption of different levels for our roads is long overdue; this is the only way in which the great traffic problem can be solved', but, in a nod to the aesthetic sensibilities of its patron, observed that 'beauty must be an inseparable companion of convenience: no longer is the merely structural or engineering solution to be accepted as sufficient'.[70] A poke in the eye for Bressey, there.

But in any case, the public reaction to the RA plans was muted at best, and critical at worst, focusing on the architecture and aesthetics rather than the functioning of London – 'style more than substance', as Larkham and Adams put it.[71] It lacked the wider focuses of planning beyond circulation. But it hardly mattered. In between the publication of the two reports, the London County Council had released what was to be Patrick Abercrombie's most significant contribution to London planning, the *County of London Plan*, and all eyes appeared to focus on that.

The same fate befell the third of Marmaras and Sutcliffe's independent plans, a 1943 publication by RIBA reporting planning activity begun in 1941.[72] As with MARS and the RA plans, transport and circulation dominated RIBA's character-ization of London's problems, probably owing to the availability of information set out in Bressey's *Highway Development Survey*, in Marmaras and Sutcliffe's view.[73] Alker Tripp popped up once more in the acknowledgements, and Raymond Unwin's work on the GLRPC attracted special praise.[74] The RIBA plan incorporated all the elements now familiar: ring roads around the centre, arterial trunk roads across the capital, full segregation between different types of traffic, the creation of zoned precincts bounded by the trunk routes, and flyover roundabouts at all main road junctions.[75] And it called for a single planning body for the region in place of decision-making by over 100 local authorities. But it had come at the wrong time. While Patrick Abercrombie was a key adviser to the RIBA committee, he was also at that time completing his *County of London Plan*, leading RIBA to pull back from their own work and back Abercrombie's official publication.

County of London Plan and *Greater London Plan*, 1943–1944

In the *County of London Plan*, authors Patrick Abercrombie and the LCC's architect, John Forshaw, were careful to widen the scope of the London planning debate beyond traffic. They presented four 'defects of present-day London', of which traffic congestion was only one (the others being housing, open spaces, and industrial zoning), and noted that 'The Highway Development Survey, 1937, was clearly a piece of planning on the strategic scale but was it sufficiently comprehensive in character, in dealing with planning as a road matter alone?' Yet even within this widened scope, of the four defects, traffic congestion was 'the most obvious and ubiquitous', leading to 'waste of time, injury and loss of life'.[76] Congestion was depicted in the *Plan* as a line of traffic on Tottenham Court Road, an image reproduced with the caption 'The traffic problem' in a Penguin paperback version of the *Plan* and shown in Figure 2.3 (attention will return to this photograph in the conclusion of this book).[77]

For Forshaw and Abercrombie, London was 'a Machine', but the machinery of movement was becoming severely overloaded.[78] The main fault lay in the mixing of through traffic with local stopping traffic and pedestrians.[79] Their solution, as has already been alluded to, was based on Alker Tripp's ideas of segregation and a hierarchy of local community areas of decreasing size, from the urban precinct to the small neighbourhood. These precincts and neighbour-hoods needed to be saved from 'invasion' by through traffic, and thus the whole ethos of the *Plan* relied on the creation of a new road network to carry the hostile invader away from the precincts and enact the geographical segregation.

Figure 2.3 'The traffic problem', *County of London Plan*, 1943–1945 (London County Council).

Peter Hall has observed that, for Abercrombie 'the roads were not merely built to reduce traffic congestion and prevent accidents; they were also to provide a new structure for London, by running in the gaps between the village and town communities'.[80] Thus, Abercrombie and Forshaw's traffic proposals followed the Bressey report 'except where its traffic routes cut through communities', in their words, and proposed a scheme to turn the whole of central London into a precinct by routing all through traffic onto a new inner 'fast motor ring-road', connected to the centre by a network of new radial and cross roads. But, unlike many earlier plans, which encouraged vertical separation through elevation, Forshaw and Abercrombie proposed to sink the new radial and cross roads in tunnels. This, they said, 'follows the lead of the tubes', and they observed that 'the motor-driver must become tunnel-minded'.[81] A further ring road, connecting the railway termini and termed a 'sub-arterial station ring', completed the plan.[82]

Frank Mort, considering visual representations expressed in the *Plan*, has noted the abstract, 'grandiose' modernity of this vision, describing its 'futuristic road diagrams' and its views from above of 'a city dominated by an abstract idea of communication'.[83] Tanis Hinchcliffe has explored its use of aerial photography.[84] This was a plan rooted in ideas of technological progress and the bringing to order of a disordered, premodern system.

It was as a master plan that the *Plan* was novel, guiding short-term decision-making so as not to prejudice longer-term opportunities. The authors expressed the customary frustration about the inability of the patchwork of authorities to plan London properly. But for the first time they were hopeful that the situation would soon be resolved by government legislation based on recent Royal Commission proposals, 'which promises to remove many of the difficulties under which planning authorities have been labouring in the past'.[85] However, they recognized that redevelopment on the scale they proposed would involve a wide consortium of both public and private enterprise.[86] There will be more on this in Chapter 7.

While working for the LCC on the *County of London Plan*, Abercrombie was also developing a plan for the Greater London region for the Ministry of Town and Country Planning. In the *Greater London Plan*, too, Abercrombie drew heavily on the 'help and inspiration' of Alker Tripp for the section on roads. Here, Abercrombie was scathing of the patchwork planning landscape, noting that amongst the 143 local authorities involved in planning in the Greater London area, 'there has been a lamentable failure to realise a need for co-ordination in planning all round London'.[87] To the four defects of London treated in the *County of London Plan* he added a fifth: urban sprawl and ribbon development, which he felt had a toxic effect on transport and congestion. The arterial roads programme of the interwar period had helped, Abercrombie observed, but government investment was woefully low, and the absence of control over the building-up of the new routes remained a huge problem. With their narrowness and the blockages caused by local traffic and pedestrians, he said, 'fast movement is thus impossible; congestion results, and accidents abound'.[88] It was the same old problem, and the *Greater London Plan* proposed the same old solution: a network of limited-access express motor roads and bypasses, with grade-separated intersections, and frontage development forbidden.[89]

City of London, 1944–1947

Abercrombie and Forshaw's work on the *County of London Plan* contained a highly significant lacuna: namely the City of London, which, as always, wished to plough its own furrow.[90] On 14 November 1940, after two months of intensive bombing by the German *Luftwaffe*, the Common Council of the City of London ordered its Committee for Improvements and Town Planning to begin work on a scheme for postwar planning and reconstruction of the Square Mile.[91] In May 1944 its preliminary report, written largely by Francis Forty, the City Engineer, was complete, being published in July. It shared with the *County* and *Greater London* plans a view that traffic congestion was caused not so much by quantity of traffic but by 'the varying attractions and social habits' of the area, which caused conflict and led to 'abnormal concentration at certain points'. It also shared the opinion that traffic intersections caused most of the problems, observing that 'Control lights and one-way streets are but palliatives to the proper solution – a well-considered street system.'[92] And, as with Abercrombie's proposals, the City plan differentiated between through and local traffic. For the former, a ring road skirting the City would take traffic away from the cramped intersection at the Bank of England.

Pedestrian subways would be provided at the ring road's roundabout connections. This had been proposed in Bressey and Lutyens's 1937 *Survey*, although the line proposed by the City Corporation differed. For local traffic, a scheme of stopping up minor streets to reduce the number of intersections was proposed, with the added advantage of aiding pedestrian movement. Finally, road layout improvements would take place at the north sides of Blackfriars, Southwark, and London Bridges. But in general, the 1944 City proposals were modest in their ambition: no overt segregation, no streets in the sky, no large-scale reconfiguration of the street plan.[93] It was, instead, a plan with 'respect for urban morphology', as Michael Hebbert has explained, in which the City Corporation 'explicitly distanced itself from the fashionable preference for clean-sweep planning', in the form of either Lutyens or Abercrombie.[94] And that was the problem.

Forty's report went down badly with both the government and critical reviewers, who felt it went nowhere near far enough.[95] Their criticism focused on Forty's status as an engineer rather than as a planner, and observed that his proposals sought simply to get the City back to the situation it was in before the war, but with worse congestion owing to increased commercial floor space, as Gordon Cherry and Leith Penny have observed.[96] One specialist noted that the report was simply a prewar transport engineer's improvement scheme for an undamaged city.[97] Soon after the report's publication in 1944, the Corporation (dictated by the government) instructed Charles Holden and William Holford to come up with something more aligned with the wider planning activity of London.

Their report was presented in 1947 (though not published until 1951) and was better received. The increased influence of Abercrombie and the LCC was clear, whether in its land-use zoning, development of local precincts, or proposals to limit the amount of accommodation in the City and therefore the working population, paralleling Abercrombian ideas of decentralization and dispersal. Holden and Holford proposed a programme of stopping-up, one-way streets, and improved intersections similar to Forty's plan, but they were more ambitious in their proposals for new roads, including west–east and north–south cross routes (one following the line of the old London Wall) and as much junction grade separation as could be fitted in.[98] Two of the new cross routes – one to the north and one to the south – were to be double-level, with the lower street taking local turning traffic and the higher level flying over existing streets. Car parking would be provided on a shelf. Overall their proposals were for a substantial intervention in the City's street land-scape (see Figure 2.4).[99]

Figure 2.4 Proposed street plan for the City of London, 1947 (detail), with edging showing new streets, building lines, and pavements (Corporation of London, courtesy of London Metropolitan Archives, City of London).

One aspect of their report was particularly novel: namely its proposal for 'pedestrian ways'. Here, Holden and Holford recognized the significance of pedestrian passage in the City, whether for business, shopping, or recreation. They presented an imaginary walk around the existing pedestrian ways of the City – in which 'people become individuals' – before proposing a wholesale linking-up of existing pedestrian routes with new connections into a circuit, 'so that it will be possible to walk around a considerable area of the City on a different plane, so to speak, from that used by the services or traffic'.[100] This was Trippian pedestrian segregation, but presented in terms of amenity, beauty, and convenience rather than safety, congestion, and fast vehicle traffic. Yet despite appearing to give moral priority to the pedestrian over the motor vehicle, Holden and Holford's pedestrian ways tended to be sunken: cars passed overhead with people beneath. Following Abercrombie, motordom was on the higher plane in a set of brutalist insertions. But this would change, as will be explored in Chapter 7.

Administrative County of London Development Plan, 1951

By the time Charles Holden and William Holford's plan for the City of London had been presented, the legislative playing field for planning in Britain had changed, with the passing of the Town and Country Planning Act of 1947, which founded a national planning system.[101] This subjected almost all development to the need for planning permission, nationalized development rights and values, and levied a charge on developers equalling the rise in land value following development. It also obliged local authorities to publish development plans for their areas projecting forward a 20-year set of proposals.[102]

For London, the planning authority was the LCC, which also took on this duty for the City of London. In 1951, the LCC published its *Administrative County of London Development Plan*. As with the plans of Abercrombie, Forshaw, Holford, and Holden during the war, the LCC plan sought to decentralize people and industry to new towns, and this was the context in which better transport links were framed. Frustrated by six years of financial pressure leaving it unable to start Abercrombie's roads programme, the LCC launched a modest manifesto of 11 miles of new main roads, 7 miles of road widening, 2 miles of new tunnels, the rebuilding of two river bridges and a river wall, and 43 new or improved traffic intersections using roundabouts or flyovers. This excluded City of London projects, the primary of which remained a new through route.[103] The priority projects for the 20-year *Development Plan* were mapped onto the LCC's 'ultimate road plan' (see Figure 2.5).[104] The ring and main roads of the Abercrombie era can be clearly seen.

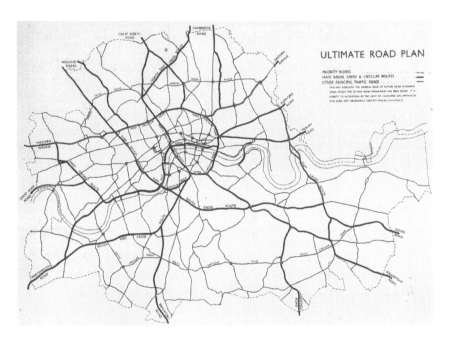

Figure 2.5 Priority road projects for the 20-year *Administrative County of London Development Plan* (1951), shown mapped onto the LCC's 'ultimate road plan' (LCC, courtesy of London Metropolitan Archives, City of London).

What was new in the *Development Plan* was hindsight of the changing levels of traffic in the capital in times of peace and war through the collection of traffic survey data. London had experienced a steady rise in road traffic in the interwar period, except for a plateau during the Depression of the early 1930s. This general rise had started to tail off as war approached, and of course the war dislocated normal traffic patterns. By 1949, traffic levels were down 10 per cent on 1937 levels, although that headline figure masked a substantial drop in private cars, heavy lorries, and public-service vehicles and a big rise in vans and taxis. Car-use was expected to pick up as supply of vehicles increased and petrol rationing had ended, but the cautious projection offered in the *Plan* was that traffic levels would only rise by 50 per cent over 1951 levels by about 1980.[105]

But Abercrombie's vision of roads in London seemed to be slowly dying, as Michael Hebbert has described.[106] In the immediate postwar period, housing had taken priority over road-building. Then, following publication of the LCC's 1951 plan, the government refused to provide financial support. And by the mid-1950s, a programme of national intercity motorways was underway. But things moved quickly in that decade and, just five years after the *Development Plan* was published, a little-known planning officer in the Ministry of Housing and Local Government issued a polemical challenge to the urban planning discipline on Britain's traffic problem, followed two years later by a popular book.[107] The stage was set for Colin Buchanan – and the crisis of the car in the capital.

Traffic in Towns, 1963

In a 1956 article, Colin Buchanan claimed that urban traffic congestion, whilst not new, seemed 'to have acquired a new urgency in the last ten years', and that it was widely held that 'unless something is done fairly soon it will be impossible for traffic to move at all'.[108] This encapsulated the view that has survived to the present: that the postwar period saw a rapid rise in urban traffic congestion that was reaching crisis levels by the turn of the 1960s.[109] Buchanan offered a new characterization of congestion, stating that 'there is far more at stake than the curing of traffic jams', preferring a definition of the word 'congestion' to mean 'all the uncivilising consequences of motor traffic in addition to obvious difficulties of movement'.[110] It was rare for a planner to define the term – rarer still for one to criticize the planning orthodoxy, which focused on enabling traffic to bypass towns. Buchanan did, describing 'a good deal of wishful thinking about the value of the by-passes'.[111]

More than any planner before him, Buchanan acknowledged the influence of Alker Tripp on the planning orthodoxy, observing that 'many of our present ideas first found coherent expression' in Tripp's 1942 book, such as traffic-limited precincts and pedestrian segregation. But Tripp's conception of precincts and the arterial roads running between them was wrong, said Buchanan. 'One begins to wonder what this traffic can possibly be that will flow so amenably along the radials, round the ring and out the other side', he remarked, concluding that 'all the evidence seems to point to a far more complicated journey-pattern than this, with far more penetration into the precincts and the central area'.[112] For Buchanan, urban congestion was mostly caused by traffic that wanted to be *in* the town, with through traffic – the target of the ring roads – being a minor problem, at least in large towns and cities such as London. He claimed 'surely what motor vehicles are mainly doing is to worm their way into alleys and courts, to shops, offices, warehouses and workshops all over the centre, and then to return more or less in their tracks'.[113] He was calling for a micro-study of traffic movement, with solutions of a fine enough grain to match, as well as a new viewpoint onto the problem in which solutions involved more than merely getting the traffic moving.

Yet on the other Tripp-influenced solution – pedestrian segregation – he was strongly in favour. The reason the idea had not seen much expression, he felt, was that the idea of a double-level urban street system 'postulates urban reconstruction on a frightening scale'. Instead, he proposed a network of elevated pedestrian walkways, cantilevered out from existing buildings along, say, Oxford Street, and connected by lightweight bridges 'thrown across at frequent intervals'. The pavements at road level would be turned over to traffic.[114]

Buchanan's iconoclastic viewpoint and focus on the micro-scale of congestion was given state sanction with the publication in 1963 of *Traffic in Towns*, written for the Ministry of Transport at the request in 1960 of Ernest Marples. Marples had become Transport Minister the previous year and had been impressed by *Mixed Blessing*, the popular book written by Buchanan to flesh out his 1956 ideas and bring them to a wider audience.[115] In *Traffic in Towns*, Buchanan described his own more sophisticated development of Tripp's precincts, which he termed 'environmental areas'.[116] Buchanan chose to explore a single 'metropolitan block' of Central London – Fitzrovia – in the hope that understanding a succession of such blocks would permit an understanding of the capital's problem as a whole – a synthetic approach diametrically at odds with the analytical approach of his predecessors.[117] But this was not to say that a focus on the local unit meant supporting the existing system of local government in urban planning; as with those that went before it, *Traffic in Towns* advocated government reform and new urban regional planning, as Simon Gunn has pointed out.[118]

Gunn has also examined the ambivalence of Buchanan's prescription, in which a concern for the local environment – his environmental areas integrated pedestrians and slowed-down vehicles – clashed with modernist proposals for fully segregated urban motorways, noting that the report was 'novel less in the ideas it contained than in their synthesis' and that whilst it did represent the end of the existing paradigm in government planning of the ring road and the circulation of traffic, nevertheless it contained a 'Corbusierian inflection' in its 'vision of multi-deck cities' among other proposals.[119]

Yet, despite the ambivalence (or perhaps because of it), the response to *Traffic in Towns* was overwhelmingly positive. Most people, it seemed, were able to buy into a vision of technological reconstruction. This was the year of Harold Wilson's 'white heat' speech on a technological revolution, and Britain's future seemed to rest on modernization. Buchanan offered that but also, cleverly, tempered it with conservationism and a focus on the small lives of the individuals. But, as both Gunn and Peter Hall have pointed out, there was one group that refused to buy his vision – the transport economists, who held that markets, not the state, should solve the problems Buchanan raised.[120] This brings up a further point raised by Buchanan back in his 1956 paper, which again had been rather heretical for an urban planner in an age of intervention, and that was the possibility of what he termed 'medicinal' as well as 'surgical' cures for traffic congestion in cities, or in other words the control of traffic through police direction; waiting restrictions; one-way routing; or, most significantly, the restriction of vehicle entry into central areas; as well as the reconstruction of the fabric of the city.

Although Buchanan's own view on such restriction was that it might be 'a useful bottle of medicine in the cupboard should the worst come to the worst', he nevertheless prophesied that it was 'in the wind' and might 'eventually play a larger part than we perhaps care to imagine at the moment'.[121] Buchanan, then, was clear that restriction rather than reconstruction could only be a palliative, and, as Gunn has shown, so were almost all contemporary responses to Buchanan's proposals except those of the economists. Both these alternative viewpoints to Buchanan's planning vision – traffic control and traffic restriction – will be discussed in later chapters. But for now, our review of plans will be concluded with one that came at a turning point in attitudes towards the car in the city, and can be seen as the final great planning statement for the holistic, state-led reconstruction of London.

Greater London Development Plan, 1969–1973 and beyond

On 28 October 1965, matters relating to congestion in London came to a head. Following several days of serious traffic jams in the evening rush hour, a three-hour emergency debate on London traffic congestion was held in the House of Commons, swiftly followed by critical leader articles in the *Evening Standard* and *Evening News*.[122] By that time, Greater London's traffic planning had been taken over by the Greater London Council (GLC), which had come into being in April that year, specifically – as Michael Hebbert has observed – to get a London road-building programme created.[123] Douglas Hart, too, has shown that the founding of the GLC enabled the traffic-planning orthodoxy of segregated limited-access motor roads, by then over 50 years old, to be enacted through its new powers combined with political will for construction.[124]

Road-building was in the ascendant on London's policy agenda. One of the GLC's first announcements – made just 12 hours after it had come into existence – had been the construction of an Inner London ring road, or 'Motorway Box', and that was just the start. A year later, in 1966, the Council announced plans for its 'Primary Road Network' and a further two motorway-standard ring roads around the capital (known, with a retitled Motorway Box, as 'Ringways') as well as a network of 12 new radial motorway routes. Then, in 1969 (two years later than expected), the GLC published its *Greater London Development Plan* (*GLDP*) which formalized the scheme.[125] In its popular version of the *GLDP*, the GLC was bullish:

> If London comes to a standstill, or even slows down to a permanent crawl, planning has failed ... Congestion means more noise, fumes, vibration ... Congestion stultifies and limits life ... Congestion is grey – grey skies, grey faces, grey fish and grey cauliflowers, grey facades and grey grass.[126]

The solution, held the GLC, was the urban motorway network; but as it turned out, the plan was doomed.

The story of the rapid decline and fall of the Ringways scheme and the Primary Road Network following a lengthy and acrimonious public inquiry has been well examined.[127] Construction of the brutal elevated Westway by the Ministry of Transport in the late 1960s showed the public the reality of urban motorways – grey, noisy, fume-laden, life-limiting. Westway opened, unfortunately for the GLC, on the eve of the *GLDP* inquiry in 1970 and became emblematic of the environmental cost of the Ringway scheme.[128] In short, by 1973, the plans for the two inner Ringways, and their associated radial routes, had become politically toxic following huge public opposition, and were pretty much taken off the table. A new paradigm of public engagement and environmental concern was becoming established, albeit fitfully – what Peter Hall has termed the 'revolt against the freeways'.[129]

And that, it seemed, was that. In 1974, J. G. Ballard wrote his fictional account of the Westway, *Concrete Island*, and the dream had officially become a nightmare.[130] The demise of the Ringways signalled the end of a planning model that had lasted three-quarters of a century, in which London's plan was conceived as a top-down skein of roads and precincts laid onto the capital. It is true that a few bits and pieces of the canonical schemes were built in the ensuing years. Parts of the East Cross Route of Ringway 1 were built as the Blackwall Tunnel Southern Approach and the Rochester Way Relief Road in the 1970s and 1980s, and a radial road connecting Ringway 1 to Ringway 2 opened after huge protests as the A12–M11 link road in 1999.[131] But there were no more grand plans for the total reconstruction of London in the twentieth century.

Yet it was not, clearly, the end of planning for roads in London. Formal planning remained after 1973, but what changed was the envisioning of roads and a moral shift in characterizations of their purpose. Public transport, cycling, and walking emerged (or re-emerged) in moral terms. Transport plans became sets of strategies. The car was still there, but reduced in status, at least temporarily. The late 1980s and the 1990s saw a new popular literature emerge on the failure of government and planning in the capital, catalysed, perhaps, by the winding-up of the GLC.[132] Changed, too, was the political landscape, both at a national and at a London level, as the capital transitioned from having no overall local government to the establishment of the Greater London Authority.[133] One crucial new approach, expressed most clearly at the London docks from about 1970 onwards, was akin to 'non-planning', an approach that appeared to herald a market free-for-all, of which the large-scale consequences on traffic policy are still playing out.[134] The concluding chapter of this book will glance at London's traffic planning landscape in the present day.

Conclusions and a view into the margins

In April 1943, the Minister of War Transport appointed a committee to look at the design and layout of roads in towns. In its 1946 report, the committee made a bold statement. 'Segregation, which may be defined as separation of traffic in the interests of safety and free flow, should be the key-note of modern road design', it stated. The solutions it put forward to achieve this urban segregation included behavioural control, civil engineering, and electrical technology as well as town planning, with ring roads, bypasses, motorways, cycle tracks, footways, one-way streets, flyovers, underpasses, and traffic-light systems all advanced as possibilities.[135]

This chapter has charted more than half-a-century of urban planning and its characterizations of traffic congestion, and has seen the themes of segregation, technology, verticality, and control emerge time after time. The peculiar patchwork of London local government has continually frustrated the attempts of planners and specialists to treat London's traffic problem holistically. There has also been a tense dialogue over who is – or should be – a planner. The sheer industry of Patrick Abercrombie is evident but, as Larkham and Adams have considered, he was very stretched.[136] As they state, 'named (and famous) plan authors were not *sole* authors'.[137] Forty at the City of London came in for criticism for being an engineer, not a planner in the disciplinary sense. But planners themselves were often architects or surveyors, at least in the earlier decades. Clearly, a wider view is needed. Here are Larkham and Adams again: 'detailed consideration should be given to the interplay, dialogue, relationships that existed between the Ministry [of Town and Country Planning], the plan authors, City officials, inhabitants and so forth, involved in shaping London's post-war future'.[138] Douglas Hart raised a similar point from a different angle, noting that:

> 'The problem' of traffic congestion was in fact several problems which varied in scope according to the values and interests of the labelling organisations. The London County Council, road-building interests, public transport authorities and various professional bodies all saw 'the problem' in different ways.[139]

Larkham and Adams's invitation to take a wider view of plans, planners, and planning is an important one.

Through numerous political statements, expressed through characterizations of London's traffic, we have a heroic narrative – that traffic is a big problem requiring a big plan (and big planners) to solve it. That there was an alternative to this was realized long before the consumer frenzy and road-building visions of the 1950s and 1960s, with motorways in the sky and a car for every citizen. The chronicler of London modernity, Harold Clunn, observed the following in 1927 as he surveyed 20 years of reconstruction:

> Wild schemes are proposed from time to time for the relief of traffic congestion in London, by the construction of overhead roads, which would cost millions of pounds ... there is little doubt that the elimination of some half a dozen of the worst 'bottlenecks' in London ... at about one-quarter the cost of the overhead road would do more to solve the traffic question than such visionary schemes as are so often suggested; but whereas these visionary projects will always receive quite an interested and respectful hearing from the Press and the general public, anything in the nature of a much less ambitious programme ... will invariably be dismissed ... as being totally inadequate to meet the requirements of the case.[140]

For Clunn, and others who came to question the need for grand interventions in London's street scene, congestion was being used by politicians to grandstand in front of voters, mediated by the press. For these politicians to be seen to solve big problems, those problems needed to be made visible – or perhaps created in the first place, since Clunn claimed that congestion in London had actually decreased over the previous 25 years. Thus the talking-up of traffic congestion by politicians eager for tangible infrastructure to open and name, fed by specialists keen to try new technologies and planning ideas, has been a feature of London for a century.

And it is a hard narrative to look beyond. Individuals such as Abercrombie and Buchanan, and artefacts such as their plans, capture our attention easily. We have thus absorbed, over many decades, such planners' conceptualizations of traffic as a means of circulation, and this has led to a common language of traffic: through routes, flow, expressways, ring roads, arterials. We have also been repeatedly exposed to big themes, including segregation and verticality, and assured that big interventions, such as the Westway, were the solution to the big problem.

But there was small-scale, fine-grain planning, too. Here, congestion was seen as local and affected individual journeys and intersections, as well as being seen in terms of aggregate flows. This predated the 1973 revolt against the freeways. Simon Gunn's recent work has explored this in Buchanan's plan, where it was most overt. Buchanan's own practice later explained that:

> 'Environmental Areas' emerged as the important complement to the Ring-way proposals: traffic and environmental management was to be used to

adapt the existing local street network in such a way as to force traffic on to a purpose designed primary and secondary road network.[141]

So by expanding this sense of planning and responding to traffic, by seeing it as more of a problem than one of circulation and flow, by examining responses to the fine grain of London traffic congestion, we can chart an alternative postwar planning history. We can ask about the histories of ideas in the canonical plans. How, for instance, did Tripp's ideas of pedestrian segregation and Buchanan's flyover decks get worked out on the ground? This was not just planning of the imagination via the pen onto the drawing board. The ideas in *Traffic in Towns* had history, in some cases tangible; they had been worked out, in part, in steel and concrete, by people in disciplines outside town planning, and in decades past, and perhaps in places far from the UK. We can ask questions about alternative discourses, or reactions to the hard interventions, such as asking who reacted against concrete reconstruction. There were people with world views who saw the hard plans as anathema, or who simply did not buy into the vision. They were urban planners, too, though not in the disciplinary sense. What did they propose? We can also ask how much of the canonical plans was put into practice before the 1973 revolt. How much trace of the road-building visions of Abercrombie, Forshaw, and Buchanan can we see on the streets of London today?

Selection of case studies

The foregoing account sets up the rationale for examining case studies across a range of professional groups and actors at the margins of professional planning. The selection of case studies, however, is a challenging one. The themes do not dictate a distinct set of cases; there are many candidates of people and projects to consider that differ from the orthodoxy. Nevertheless, some assessment criteria can be proposed in order to lead towards a rich and nuanced view as well as to attempt to offset biases or prejudices over what constitutes the traffic problem – biases and prejudices that are the very concern of this book. The criteria against which candidate case studies in the book have been assessed are as follows.

The first is a level hierarchy of actors. This has meant looking for middle-ranking officials as well as the Abercrombies and Buchanans, and this is designed to offset the singular heroic trope. The second criterion is geographical location (seeking a range from west to east, central to outer London) and land-use type (residential, industrial, public, retail). This throws up a range of traffic types, from freight haulage to commuting, and from leisure travel to local shopping. Together, this spread of locations and traffic types is meant to hedge broadly against biases of socio-economic status, as well as, to a certain extent, gender and age. The third criterion is chronological, looking for coverage across the period of the twentieth century. A fourth consideration is the scale of each project under review to ensure small-scale interventions are examined as well as those covering the whole city. At its least this means looking for mundane technologies as well as exceptional ones, and hedges against an obsession with size and form. A fifth concern underpins the other four. This is about acting on responses developed by hands-on engagement with the subject – learning by walking the actual streets of London. The intention has been to take a heuristic, instinctive approach to spotting good candidate case studies lurking in the literature. With these problems, themes, questions, and guidelines in mind, the rationale for the selection of case studies, and the ways they allow us to study the margins of the network, is as follows.

Chapter 3, which considers congestion as an engineering problem, explores the construction of the 1934 Silvertown Viaduct in east London. By doing so, we can place ideas expressed a short time later by Charles Bressey in the *Highway Development Survey* into a context wider than town planning. These include an engineering world view of efficiency, in part driven by the affordances of the new materials and techniques of steel-reinforced concrete, as well as a political vision strongly inflected by class politics of labour and trade over a crucial period in British and international socio-political history. As a project at the geographical and social margins of east London, designed primarily for freight vehicles and use by working-class people, it provides an alternative narrative on interwar road-building modernities in the capital, which have tended to focus on arterial roads, automobility, and the affluent middle classes.

Chapter 4, which looks at congestion as a problem of control, examines the concept of segregation, an idea embedded in professional plans for London's traffic network. Metropolitan Police Assistant Commissioner Alker Tripp sits in the margins of the planning literature as an influence on this concept, yet, apart from noting that his police work gave him a deep insight into the problems of traffic and the dangers caused by traffic collisions, little has been said to unpack the relationship between his work as a cop and his approach as a planner. He is therefore at once a canonical figure and a marginal voice.

Chapter 5, which examines a technological alternative to the planners' reconstructionist paradigm, namely road pricing, emerged from a close reading of recent literature around Colin Buchanan's *Traffic in Towns*, which has observed that it was only economists who had any real problem with Buchanan's proposals, suggesting instead a market mechanism for allocation of space mediated through pricing. We have a discipline – economics – that was *actively* marginalized from the planning discourse by planning professionals.

Chapter 6 looks at a 1960s west London experiment in computer-controlled traffic lights. Traffic lights and roadway vehicle sensors exist in the margins of our vision. Few of us ever notice the rectangular patterns of poured tar near stop-lines, or cast-iron boxes by the side of the road. It has been easy, therefore, to ignore the traffic-light network growing since the 1960s, in favour of studying the Ringways and similar big structures. But by examining the electronic network, one starts to get a sense of the physical as well as the socio-political scale of this infrastructure. And yet we ignore it. Chapter 6 will offer suggestions as to why we should not.

Chapter 7 looks at three cases offering glimpses into traffic and the mobilities of capital. The first and second – Centre Point at St Giles' Circus and the London Wall highway – have already received close attention, and are discussed here to explore the ways planners and developers attempted to exploit each other to gain beneficial outcomes. However, the third study – the modest traffic calming of the Pimlico Precinct – has received little attention. Yet, as the activist and local press observed at the time, the relationship between this traffic project and wider issues of property capitalism is significant, and helps reveal the long history of traffic projects saturated with market decisions and the accumulation of capital.

Notes

1 Hermione Hobhouse, 'The First Real Move to a Planning of London', *Urban Design Quarterly* (June 1989), 19.
2 Gordon Cherry, *The Evolution of British Town Planning* (Leighton Buzzard: Leonard Hill, 1974), 107, 125–7, 193–6.
3 Donald Foley, *Controlling London's Growth: Planning the Great Wen 1940–1960* (Berkeley: University of California Press, 1963), 111–12.
4 Helen Meller, *Towns, Plans and Society in Modern Britain* (Cambridge: Cambridge University Press, 1997), 82.
5 Douglas Hart, *Strategic Planning in London: The Rise and Fall of the Primary Road Network* (Oxford: Pergamon Press, 1976), 26.
6 Peter Larkham and David Adams, *The Post-War Reconstruction Planning of London: A Wider Perspective*, Working Paper Series 8 (Birmingham: Centre for Environment and Society Research, Birmingham City University, 2011), 1.
7 James Winter, *London's Teeming Streets, 1830–1914* (London: Routledge, 1993).
8 Meller.
9 Ebenezer Howard, *Tomorrow: A Peaceful Path to Real Reform* (London: Son-nenschein, 1898); see also John Rockey, 'From Vision to Reality: Victorian Ideal Cities and Model Towns in the Genesis of Ebenezer Howard's Garden City', *Town Planning Review* 54, no. 1 (January 1983): 83–105; Peter Hall, *Cities of Tomorrow* (Oxford: Blackwell, 1988), Chapter 4; Anthony Sutcliffe, 'From Town–Country to Town Planning: Changing Priorities in the British Garden City Movement, 1899–1914', *Planning Perspectives* 5 (1990): 257–69.
10 Cherry.
11 Hall, *Cities of Tomorrow*, Chapter 6; Stephen Ward, *Planning the Twentieth-Century City: The Advanced Capitalist World* (Chichester: John Wiley & Sons, 2002), 35–7; Joseph DiMento and Cliff Ellis, *Changing Lanes: Visions and Histories of Urban Freeways* (Cambridge, MA: MIT Press, 2013), 10–12.
12 Jane Jacobs, *The Death and Life of Great American Cities: The Failure of Town Planning* (London: Pelican, 1965 [1961]), 35.
13 Hall, *Cities of Tomorrow*, Chapter 7; Ward, *Planning the Twentieth-Century City*, 98–100.
14 Jacobs, 33.
15 Timothy Davis, '"A pleasant illusion of unspoiled countryside": The American Parkway and the Problematics of an Institutionalized Vernacular', *Perspectives in Vernacular Architecture* 9 (2003): 228–46; Thomas Zeller, 'Staging the Driving Experience: Parkways in Germany and the United States', in *Routes, Roads and*

Landscapes, ed. Mari Hvattum, Brita Brenna, Beate Elvebakk, and Janike Kampe-vold Larsen (Farnham: Ashgate, 2011), 125–38.

16 Carmen Hass-Klau, *The Pedestrian and City Traffic* (London: Belhaven Press, 1990), Chapter 2.

17 Helena Beaufoy, '"Order out of chaos": The London Society and the Planning of London 1912–1920', *Planning Perspectives* 12 (1997): 135–64.

18 Lucy Hewitt, 'The London Society and Their Development Plan for Greater London', *London Topographical Record* 30 (2010): 115–31; Lucy Hewitt, 'Towards a Greater Urban Geography: Regional Planning and Associational Networks in London during the Early Twentieth Century', *Planning Perspectives* 26, no. 4 (October 2011): 551–68.

19 Royal Institute of British Architects, *Town Planning Conference, London, 10–15 October 1910: Transactions* (London: Royal Institute of British Architects, 1911); Hewitt, 'The London Society and Their Development Plan for Greater London', 116; see also Fabiano Lemes De Oliveira, 'The First Town Planning Conference Revisited (1910–2010)', http://eprints.port.ac.uk/6306/.

20 Hewitt, 'The London Society and Their Development Plan for Greater London', 120.

21 On the arterial roads programme, see for instance Cherry, 69–70; and Michael John Law, *The Experience of Suburban Modernity: How Private Transport Changed Interwar London* (Manchester: Manchester University Press, 2014), Chapter 5.

22 David Gilbert, 'London of the Future: The Metropolis Reimagined after the Great War', *Journal of British Studies* 43, no. 1 (January 2004): 97.

23 London Society, *Development Plan of Greater London* (London: E. Stanford, 1919).

24 Aston Webb, 'The London Society's Map, with Its Proposals for the Improvement of London', *Geographical Journal* 51, no. 5 (May 1918): 273–93; London Society, *London of the Future*, ed. Aston Webb (London: T. Fisher Unwin, 1921), 19.

25 Reproduced in Hewitt, 'The London Society and Their Development Plan for Greater London', 122.

26 Webb, 273–9.

27 Gilbert, 101; on decentralization see for instance David Rooney, 'Visualization, Decentralization and Metropolitan Improvement: "Light-and-Air" and London County Council Photographs, 1899–1908', *Urban History* 40, no. 3 (August 2013): 462–82.

28 Hewitt, 'The London Society and Their Development Plan for Greater London', 121–3.

29 Charles Bressey and Edwin Lutyens, *Ministry of Transport Highway Development Survey 1937 (Greater London)* (London: HMSO, 1938), 24.

30 Hewitt, 'The London Society and Their Development Plan for Greater London', 125–6.

31 London Society, *London of the Future*.

32 Gilbert, 92.

33 Gilbert, 94.

34 Hewitt, 'The London Society and Their Development Plan for Greater London', 128–9.

35 Hewitt, 'The London Society and Their Development Plan for Greater London', 121.

36 Raymond Unwin, 'Some Thoughts on the Development of London', in London Society, *London of the Future*, 177–92; Gilbert, 101–2.

37 Greater London Regional Planning Committee, *First Report* (London: Knapp, Drewett, & Sons, 1929); Greater London Regional Planning Committee, *Second Report* (London: Knapp, Drewett, & Sons, 1933).

38 Clough Williams-Ellis, *England and the Octopus* (London: Geoffrey Bles, 1928), 161–6.

39 Greater London Regional Planning Committee, *First Report*, 27–9.

40 Greater London Regional Planning Committee, *Second Report*, 62.
41 Greater London Regional Planning Committee, *Second Report*, 63–9.
42 Greater London Regional Planning Committee, *Second Report*, 70–3.
43 Rees Jeffreys, *The King's Highway: An Historical and Autobiographical Record of the Developments of the Past Sixty Years* (London: Batchworth Press, 1949), 222.
44 Bressey and Lutyens, 5; Charles Bressey, 'Greater London Highway Development Survey: Discussion', *Geographical Journal* 94, no. 5 (November 1939): 361.
45 Royal Commission on London Traffic, *Report*, Vol. I (London: HMSO, 1905).
46 Bressey and Lutyens, 5.
47 Bressey and Lutyens, 11–12. The junctions were as follows: Oxford Circus; Gardiner's Corner, Aldgate; High Holborn/Kingsway; Hart Street/Southampton Row/Theobalds Road; Hammersmith Broadway; Kennington Triangle; Angel, Islington; Highgate Hill/Archway Road; St Johns Road/Holloway Road/Junction Road; Cambridge Circus; Camberwell Green; Mornington Crescent Station intersection; Britannia, Camden Town intersection; and the Elephant and Castle.
48 Bressey and Lutyens, end pocket.
49 Bressey and Lutyens, 28–9.
50 Bressey and Lutyens, 17–18.
51 Bressey and Lutyens, 34.
52 Bressey and Lutyens, 6, 34.
53 See for instance William Ashworth, *The Genesis of Modern British Town Planning* (London: Routledge & Kegan Paul, 1954), 216; and Cherry, 107.
54 Patrick Aberbcrombie in Alker Tripp, *Town Planning and Road Traffic* (London: E. Arnold, 1942); Peter Hall, *Great Planning Disasters* (Harmondsworth: Penguin, 1980), 59; Peter Hall, 'Bringing Abercrombie back from the Shades: A Look Forward and Back', *Town Planning Review* 66, no. 3 (July 1995): 232; Stephen Ward, *Planning and Urban Change*, 2nd edn (London: Sage, 2004), 88; J. A. Proudlove, 'Transport Planning', *Town Planning Review* 39, no. 2 (July 1968): 87; anon. [David Massey?], 'Leslie Patrick Abercrombie: A Centenary Note', *Town Planning Review* 50, no. 3 (July 1979): 258.
55 Alker Tripp, *Road Traffic and Its Control* (London: E. Arnold, 1938).
56 Peter Hall, *London 2000* (London: Faber and Faber, 1963), 208–9; Hall, *Great Planning Disasters*, 64; Hart, 97; Hass-Klau, Chapter 8; Ward, *Planning and Urban Change*, 135–6; Simon Gunn, 'The Buchanan Report, Environment and the Problem of Traffic in 1960s Britain', *Twentieth Century British History* 22, no. 4 (2011): 523.
57 Emmanuel Marmaras and Anthony Sutcliffe, 'Planning for Post-War London: The Three Independent Plans, 1942–3', *Planning Perspectives* 9, no. 4 (1994): 431–53.
58 Charlotte Benton, 'Korn, Arthur (1891–1978)', *Oxford Dictionary of National Biography* (2004); Frank Newby, 'Samuely, Felix James (1902–1959)', *Oxford Dictionary of National Biography (Online Edition)* (September 2010).
59 Arthur Korn and Felix J. Samuely, 'A Master Plan for London: Based on the Research Carried Out by the Town Planning Committee of the M.A.R.S. Group', *Architectural Review* 91 (1942): 143–51; see also Arthur Korn, Maxwell Fry, and Dennis Sharp, 'The M.A.R.S. Plan for London', *Perspecta* 13/14 (1971): 163–73.
60 Marmaras and Sutcliffe, 434. On the origin of some of MARS's transport ideas, and wider contexts of the MARS plan, see John Gold, 'Towards the Functional City? MARS, CIAM and the London Plans 1933–42', in *The Modern City Revisited*, ed. Thomas Deckker (London: Spon Press, 2000), 80–99.
61 Marmaras and Sutcliffe, 434–5.
62 Korn and Samuely, 144.
63 Gold, 92.
64 Korn and Samuely, 147.
65 Gold, 81.

66 Royal Academy Planning Committee, *London Replanned: The Royal Academy Planning Committee's Interim Report* (London: Country Life, 1942); Royal Academy Planning Committee, *Road, Rail and River in London: The Royal Academy Planning Committee's Second Report* (London: Country Life, 1944).

67 Marmaras and Sutcliffe, 440–1.

68 Marmaras and Sutcliffe, 443; Lionel Esher, *A Broken Wave* (London: Viking, 1981), 95.

69 Royal Academy Planning Committee, *Road, Rail and River in London*, 4.

70 Royal Academy Planning Committee, *Road, Rail and River in London*, 6.

71 Larkham and Adams, 24.

72 Royal Institute of British Architects, *Greater London: Towards a Master Plan (Second Interim Report of the London Regional Reconstruction Committee)* (London: Royal Institute of British Architects, 1943).

73 Marmaras and Sutcliffe, 447.

74 Royal Institute of British Architects, *Greater London*, 21.

75 Marmaras and Sutcliffe, 444–5.

76 John Forshaw and Patrick Abercrombie, *County of London Plan* (London: London County Council/Macmillan and Co., 1943), 2–3.

77 E. J. Carter and Ernö Goldfinger, *The County of London Plan: Explained by E. J. Carter and Ernö Goldfinger* (London: Penguin, 1945), 9; originally published (without title and with slightly wider view) in Forshaw and Abercrombie, facing 8.

78 Forshaw and Abercrombie, 7.

79 Forshaw and Abercrombie, 1–4.

80 Hall, 'Bringing Abercrombie back from the Shades', 232.

81 Forshaw and Abercrombie, 13, 10–11.

82 Forshaw and Abercrombie, 14. The authors noted that the North and South Orbital and Circular ring roads were already planned and partly executed, though outside the scope of their area. In all, four ring roads were envisaged for the capital.

83 Frank Mort, 'Fantasies of Metropolitan Life: Planning London in the 1940s', *Journal of British Studies* 43, no. 1 (January 2004): 144.

84 Tanis Hinchcliffe, 'Aerial Photography and the Postwar Urban Planner in London', *London Journal* 35, no. 3 (November 2010): 277–88.

85 Forshaw and Abercrombie, 49.

86 Forshaw and Abercrombie, 17–20.

87 Patrick Abercrombie, *Greater London Plan 1944* (London: HMSO, 1945), iv.

88 Abercrombie, 4.

89 Abercrombie, 9–10.

90 Larkham and Adams, 17.

91 On the City of London's planning activity, and its relationships with other plans, see Junichi Hasegawa, 'Governments, Consultants and Expert Bodies in the Physical Reconstruction of the City of London in the 1940s', *Planning Perspectives* 14 (1999): 121–44.

92 Corporation of London Improvements and Town Planning Committee, *Reconstruction in the City of London* (London: Corporation of London, 1944), 19.

93 Corporation of London Improvements and Town Planning Committee, *Reconstruction in the City of London*, 19–27.

94 Michael Hebbert, *London: More by Fortune than Design* (Chichester: John Wiley & Sons, 1998), 70.

95 Hebbert, *London*, 71–2; Larkham and Adams, 17–20.

96 Gordon Cherry and Leith Penny, *Holford: A Study in Architecture, Planning and Civic Design* (London: Mansell, 1986), 136.

97 Larkham and Adams, 18.

98 Corporation of London Improvements and Town Planning Committee, *The City of London: A Record of Destruction and Survival* (London: Architectural Press for the Corporation of London, 1951), 56–65; Cherry and Penny, 137–8.

99 Corporation of London Improvements and Town Planning Committee, *The City of London*, facing 64.

100 Corporation of London Improvements and Town Planning Committee, *The City of London*, 72, 232.

101 Cherry, 144–5.

102 Stephen Elkin, *Politics and Land Use Planning* (Cambridge: Cambridge University Press, 1974), 23.

103 London County Council, *Administrative County of London Development Plan 1951: Analysis* (London: London County Council, 1951), 3.

104 London County Council, 160.

105 London County Council, 142–3.

106 Hebbert, *London*, Chapter 4.

107 Colin Buchanan, 'The Road Traffic Problem in Britain', *Town Planning Review* 26, no. 4 (January 1956): 215–37; Colin Buchanan, *Mixed Blessing: The Motor in Britain* (London: L. Hill, 1958).

108 Buchanan, 'The Road Traffic Problem in Britain', 218.

109 Gunn, 523–24.

110 Buchanan, 'The Road Traffic Problem in Britain', 222.

111 Buchanan, 'The Road Traffic Problem in Britain', 221.

112 Buchanan, 'The Road Traffic Problem in Britain', 227.

113 Buchanan, 'The Road Traffic Problem in Britain', 225.

114 Buchanan, 'The Road Traffic Problem in Britain', 235.

115 Hass-Klau, 159; Gunn, 525.

116 Ministry of Transport, *Traffic in Towns: A Study of the Long Term Problems of Traffic in Urban Areas* (London: HMSO, 1963), Chapter 2.

117 Ministry of Transport, *Traffic in Towns*, 124–64; see also Hart, 93.

118 Gunn, 531.

119 Gunn, 528–9.

120 Gunn, 531–3; Peter Hall, 'The Buchanan Report: 40 Years On', *Proceedings of the Institution of Civil Engineers: Transport* 157, no. 1 (February 2004): 7–14.

121 Buchanan, 'The Road Traffic Problem in Britain', 236–7.

122 Hart, 29–31.

123 Hebbert, *London*, 76.

124 Hart, Chapter 4.

125 Greater London Council, *Greater London Development Plan: Statement* (London: Greater London Council, 1969); Greater London Council, *Greater London Development Plan: Report of Studies* (London: Greater London Council, 1969); Greater London Council, *Tomorrow's London: A Background to the Greater London Development Plan* (London: Greater London Council, 1969); on the lateness of the *Plan* see Donald Hagman, 'The Greater London Development Plan Inquiry', *Journal of the American Institute of Planners* 37, no. 5 (1971): 290.

126 Greater London Council, 62.

127 Hart, Chapter 5; Hall, *Great Planning Disasters*, Chapter 3; David Starkie, *The Motorway Age: Road and Traffic Policies in Post-War Britain* (Oxford: Pergamon Press, 1982), Chapters 8–9; Hagman; Hebbert, 76–7; Gunn, 541–2; and, for a partisan view at the time from within the opposition camp, see J. Michael Thomson, *Motorways in London* (London: Gerald Duckworth, 1969).

128 Starkie, 77; Kathryn A. Morrison and John Minnis, *Carscapes: The Motor Car, Architecture and Landscape in England* (New Haven: Yale University Press, 2012), 356–7.

129   Hall, *Cities of Tomorrow*, 317.
130   J. G. Ballard, *Concrete Island* (London: Jonathan Cape, 1974).
131   The Rochester Way Relief Road project in the aftermath of the abolition of the GLC
      is touched on in Michael Hebbert and Ann Dickins Edge, *Dismantlers: The London
      Residuary Body* (London: Greater London Group, 1994), 119–20. On the A12–M11
      link road see *New Civil Engineer*, 12 November 1998, 24–6; and Sir Peter Baldwin,
      Robert Baldwin, and Dewi Ieuan Evans, eds., *The Motorway Achievement: Building
      the Network in Southern and Eastern England* (Chichester: Phillimore for the
      Motorway Archive Trust, 2007), 137–9. For detailed information about the Ring-
      ways and other urban motorway schemes, see the CBRD and SABRE websites at
      www.cbrd.co.uk and www.sabre-roads.org.uk.
132   See for instance the essays in Andy Thornley, ed., *The Crisis of London* (London:
      Routledge, 1992); and Edward Platt, *Leadville: A Biography of the A40* (London:
      Picador, 2000).
133   Hebbert; Tony Travers, *The Politics of London: Governing an Ungovernable City*
      (Basingstoke: Palgrave Macmillan, 2004).
134   Reyner Banham, Paul Barker, Peter Hall, and Cedric Price, 'Non-Plan: An Experi-
      ment in Freedom', *New Society* 338 (20 March 1969): 435–43; Michael Hebbert,
      'One "Planning Disaster" after Another: London Docklands 1970–1992', *The
      London Journal* 17, no. 2 (1992): 115–34; Henry Miller, *London Docklands Devel-
      opment Corporation Research Paper: An Introductory Briefing* (London: Eastside
      Community Heritage, 2009).
135   Ministry of Transport, *Design and Layout of Roads in Built-Up Areas* (London:
      HMSO, 1946), 44.
136   Larkham and Adams, *The Post-War Reconstruction Planning of London*, 31–3.
137   Larkham and Adams, 35.
138   Larkham and Adams, 36.
139   Hart, *Strategic Planning in London*, 26–7.
140   Harold Clunn, *London Rebuilt 1897–1927* (London: John Murray, 1927), 6–7.
141   Malcolm Buchanan, Nicholas Bursey, Kingsley Lewis, and Paul Mullen, *Transport
      Planning for Greater London* (Hampshire: Saxon House, 1980), 266.

Bibliography

New Civil Engineer, November 1998.

Abercrombie, Patrick. *Greater London Plan 1944*. London: HMSO, 1945.

Anon. [David Massey?]. 'Leslie Patrick Abercrombie: A Centenary Note'. *Town Planning
   Review* 50, no. 3 (July 1979): 257–64.

Ashworth, William. *The Genesis of Modern British Town Planning*. London: Routledge &
   Kegan Paul, 1954.

Baldwin, Sir Peter, Robert Baldwin, and Dewi Ieuan Evans, eds. *The Motorway Achievement: Building the Network in Southern and Eastern England*. Chichester: Phillimore for
   the Motorway Archive Trust, 2007.

Ballard, J. G. *Concrete Island*. London: Jonathan Cape, 1974.

Banham, Reyner, Paul Barker, Peter Hall, and Cedric Price. 'Non-Plan: An Experiment in
   Freedom'. *New Society* 338 (20 March 1969): 435–43.

Beaufoy, Helena. '"Order out of chaos": The London Society and the Planning of London
   1912–1920'. *Planning Perspectives* 12 (1997): 135–64.

Benton, Charlotte. 'Korn, Arthur (1891–1978)'. *Oxford Dictionary of National Biography*,
   2004.

Bressey, Charles. 'Greater London Highway Development Survey: Discussion'. *Geographical Journal* 94, no. 5 (November 1939): 353–67.

Bressey, Charles, and Edwin Lutyens. *Ministry of Transport Highway Development Survey 1937 (Greater London)*. London: HMSO, 1938.

Buchanan, Colin. *Mixed Blessing: The Motor in Britain*. London: L. Hill, 1958.

———. 'The Road Traffic Problem in Britain'. *The Town Planning Review* 26, no. 4 (January 1956): 215–37.

Buchanan, Malcolm, Nicholas Bursey, Kingsley Lewis, and Paul Mullen. *Transport Planning for Greater London*. Hampshire: Saxon House, 1980.

Carter, E. J., and Ernö Goldfinger. *The County of London Plan: Explained by E. J. Carter and Ernö Goldfinger*. London: Penguin, 1945.

Cherry, Gordon. *The Evolution of British Town Planning*. Leighton Buzzard: Leonard Hill, 1974.

Cherry, Gordon, and Leith Penny. *Holford: A Study in Architecture, Planning and Civic Design*. London: Mansell, 1986.

Clunn, Harold. *London Rebuilt 1897–1927*. London: John Murray, 1927.

Corporation of London Improvements and Town Planning Committee. *The City of London: A Record of Destruction and Survival*. London: Architectural Press for the Corporation of London, 1951.

———. *Reconstruction in the City of London*. London: Corporation of London, 1944.

Davis, Timothy. '"A pleasant illusion of unspoiled countryside": The American Parkway and the Problematics of an Institutionalized Vernacular'. *Perspectives in Vernacular Architecture* 9 (2003): 228–46.

DiMento, Joseph, and Cliff Ellis. *Changing Lanes: Visions and Histories of Urban Freeways*. Cambridge, MA: MIT Press, 2013.

Elkin, Stephen. *Politics and Land Use Planning*. Cambridge: Cambridge University Press, 1974.

Esher, Lionel. *A Broken Wave*. London: Viking, 1981.

Foley, Donald. *Controlling London's Growth: Planning the Great Wen 1940–1960*. Berkeley: University of California Press, 1963.

Forshaw, John, and Patrick Abercrombie. *County of London Plan*. London: London County Council/Macmillan and Co., 1943.

Gilbert, David. 'London of the Future: The Metropolis Reimagined after the Great War'. *Journal of British Studies* 43, no. 1 (January 2004): 91–119, 163–6.

Gold, John. 'Towards the Functional City? MARS, CIAM and the London Plans 1933–42'. In *The Modern City Revisited*, ed. Thomas Deckker, 80–99. London: Spon Press, 2000.

Greater London Council. *Greater London Development Plan: Report of Studies*. London: Greater London Council, 1969.

———. *Greater London Development Plan: Statement*. London: Greater London Council, 1969.

———. *Tomorrow's London: A Background to the Greater London Development Plan*. London: Greater London Council, 1969.

Greater London Regional Planning Committee. *First Report*. London: Knapp, Drewett, & Sons, 1929.

———. *Second Report*. London: Knapp, Drewett, & Sons, 1933.

Gunn, Simon. 'The Buchanan Report, Environment and the Problem of Traffic in 1960s Britain'. *Twentieth Century British History* 22, no. 4 (2011): 521–42.

Hagman, Donald. 'The Greater London Development Plan Inquiry'. *Journal of the American Institute of Planners* 37, no. 5 (1971): 290–96.

Hall, Peter. 'Bringing Abercrombie back from the Shades: A Look Forward and Back'. *Town Planning Review* 66, no. 3 (July 1995): 227–41.

———. 'The Buchanan Report: 40 Years On'. *Proceedings of the Institution of Civil Engineers: Transport* 157, no. 1 (February 2004): 7–14.

———. *Cities of Tomorrow.* Oxford: Blackwell, 1988.

———. *Great Planning Disasters.* Harmondsworth: Penguin, 1980.

———. *London 2000.* London: Faber and Faber, 1963.

Hart, Douglas. *Strategic Planning in London: The Rise and Fall of the Primary Road Network.* Oxford: Pergamon Press, 1976.

Hasegawa, Junichi. 'Governments, Consultants and Expert Bodies in the Physical Reconstruction of the City of London in the 1940s'. *Planning Perspectives* 14 (1999): 121–44.

Hass-Klau, Carmen. *The Pedestrian and City Traffic.* London: Belhaven Press, 1990.

Hebbert, Michael. *London: More by Fortune than Design.* Chichester: John Wiley & Sons, 1998.

———. 'One "Planning Disaster" after Another: London Docklands 1970–1992'. *London Journal* 17, no. 2 (1992): 115–34.

Hebbert, Michael, and Ann Dickins Edge. *Dismantlers: The London Residuary Body.* London: Greater London Group, 1994.

Hewitt, Lucy. 'The London Society and Their Development Plan for Greater London'. *London Topographical Record* 30 (2010): 115–31.

———. 'Towards a Greater Urban Geography: Regional Planning and Associational Networks in London during the Early Twentieth Century'. *Planning Perspectives* 26, no. 4 (October 2011): 551–68.

Hinchcliffe, Tanis. 'Aerial Photography and the Postwar Urban Planner in London'. *London Journal* 35, no. 3 (November 2010): 277–88.

Hobhouse, Hermione. 'The First Real Move to a Planning of London'. *Urban Design Quarterly* (June 1989): 17–21.

Howard, Ebenezer. *Tomorrow: A Peaceful Path to Real Reform.* London: Sonnenschein, 1898.

Jacobs, Jane. *The Death and Life of Great American Cities: The Failure of Town Planning.* London: Pelican, 1965 [1961].

Jeffreys, Rees. *The King's Highway: An Historical and Autobiographical Record of the Developments of the Past Sixty Years.* London: Batchworth Press, 1949.

Korn, Arthur, and Felix J. Samuely. 'A Master Plan for London: Based on the Research Carried Out by the Town Planning Committee of the M.A.R.S. Group'. *Architectural Review* 91 (1942): 143–51.

Korn, Arthur, Maxwell Fry, and Dennis Sharp. 'The M.A.R.S. Plan for London'. *Perspecta* 13/14 (1971): 163–73.

Larkham, Peter, and David Adams. *The Post-War Reconstruction Planning of London: A Wider Perspective.* Working Paper Series 8. Birmingham: Centre for Environment and Society Research, Birmingham City University, 2011.

Law, Michael John. *The Experience of Suburban Modernity: How Private Transport Changed Interwar London.* Manchester: Manchester University Press, 2014.

Lemes De Oliveira, Fabiano. 'The First Town Planning Conference Revisited (1910–2010)'. http://eprints.port.ac.uk/6306/.

London County Council. *Administrative County of London Development Plan 1951: Analysis.* London: London County Council, 1951.

London Society. *Development Plan of Greater London.* London: E. Stanford, 1919.

———. *London of the Future*, ed. Aston Webb. London: T. Fisher Unwin, 1921.

Marmaras, Emmanuel, and Anthony Sutcliffe. 'Planning for Post-War London: The Three Independent Plans, 1942–3'. *Planning Perspectives* 9, no. 4 (1994): 431–53.

Meller, Helen. *Towns, Plans and Society in Modern Britain.* Cambridge: Cambridge University Press, 1997.

Miller, Henry. *London Docklands Development Corporation Research Paper: An Introductory Briefing.* London: Eastside Community Heritage, 2009.

Ministry of Transport. *Design and Layout of Roads in Built-Up Areas.* London: HMSO, 1946.

———. *Traffic in Towns: A Study of the Long Term Problems of Traffic in Urban Areas.* London: HMSO, 1963.

Morrison, Kathryn A., and John Minnis. *Carscapes: The Motor Car, Architecture and Landscape in England.* New Haven: Yale University Press, 2012.

Mort, Frank. 'Fantasies of Metropolitan Life: Planning London in the 1940s'. *Journal of British Studies* 43, no. 1 (January 2004): 120–51, 167–72.

Newby, Frank. 'Samuely, Felix James (1902–1959)'. *Oxford Dictionary of National Biography (Online Edition)* (September 2010).

Platt, Edward. *Leadville: A Biography of the A40.* London: Picador, 2000.

Proudlove, J. A. 'Transport Planning'. *Town Planning Review* 39, no. 2 (July 1968): 85–98.

Rockey, John. 'From Vision to Reality: Victorian Ideal Cities and Model Towns in the Genesis of Ebenezer Howard's Garden City'. *Town Planning Review* 54, no. 1 (January 1983): 83–105.

Rooney, David. 'Visualization, Decentralization and Metropolitan Improvement: "Light-and-Air" and London County Council Photographs, 1899–1908'. *Urban History* 40, no. 3 (August 2013): 462–82.

Royal Academy Planning Committee. *London Replanned: The Royal Academy Planning Committee's Interim Report.* London: Country Life, 1942.

———. *Road, Rail and River in London: The Royal Academy Planning Committee's Second Report.* London: Country Life, 1944.

Royal Commission on London Traffic. *Report.* Vol. I. London: HMSO, 1905.

Royal Institute of British Architects. *Greater London: Towards a Master Plan (Second Interim Report of the London Regional Reconstruction Committee).* London: Royal Institute of British Architects, 1943.

———. *Town Planning Conference, London, 10–15 October 1910: Transactions.* London: Royal Institute of British Architects, 1911.

Starkie, David. *The Motorway Age: Road and Traffic Policies in Post-War Britain.* Oxford: Pergamon Press, 1982.

Sutcliffe, Anthony. 'From Town–Country to Town Planning: Changing Priorities in the British Garden City Movement, 1899–1914'. *Planning Perspectives* 5 (1990): 257–69.

Thomson, J. Michael. *Motorways in London.* London: Gerald Duckworth, 1969.

Thornley, Andy, ed. *The Crisis of London.* London: Routledge, 1992.

Travers, Tony. *The Politics of London: Governing an Ungovernable City.* Basingstoke: Palgrave Macmillan, 2004.

Tripp, Alker. *Road Traffic and Its Control.* London: E. Arnold, 1938.

———. *Town Planning and Road Traffic.* London: E. Arnold, 1942.

Unwin, Raymond. 'Some Thoughts on the Development of London'. In London Society, *London of the Future*, ed. Aston Webb, 177–92. London: T. Fisher Unwin, 1921.

Ward, Stephen. *Planning and Urban Change.* 2nd edn. London: Sage, 2004.

———. *Planning the Twentieth-Century City: The Advanced Capitalist World.* Chichester: John Wiley & Sons, 2002.

Webb, Aston. 'The London Society's Map, with Its Proposals for the Improvement of London'. *Geographical Journal* 51, no. 5 (May 1918): 273–93.

Williams-Ellis, Clough. *England and the Octopus*. London: Geoffrey Bles, 1928.

Winter, James. *London's Teeming Streets, 1830–1914*. London: Routledge, 1993.

Zeller, Thomas. 'Staging the Driving Experience: Parkways in Germany and the United States'. In *Routes, Roads and Landscapes*, ed. Mari Hvattum, Brita Brenna, Beate Elvebakk, and Janike Kampevold Larsen, 125–38. Farnham: Ashgate, 2011.

3 Engineers, flyovers, and empires

Introduction

In September 1934, a new road was opened in Canning Town, east London, leading to the Royal Victoria Dock. Named 'Silvertown Way', this was no ordinary road. It was a reinforced-concrete elevated highway described as 'a remarkable engineering feat', which rose over the existing network of roads, railways, and waterways in an area formerly 'half-strangled in a maze of narrow approaches'.[1] This was Britain's first modern urban flyover, a lofty structure sitting on 6,000 concrete piles sunk deep into the ground, carrying goods and workers to and from the heart of London's docks.

It was an engineers' solution to traffic congestion in which new roads passed over old ones, a grade-separated city that abolished conflict and fostered healthy circulation, and was the brainchild of two of Britain's most influential highway engineers, namely Henry Maybury and his protégé, Charles Bressey, at the Ministry of Transport. These two engineers profoundly shaped London's road-scape, partly during their own careers, in projects such as Silvertown Way and London's arterial road programme, and partly in the influence they exerted on a generation of urban planners who were waiting in the wings ready to act during and after the Second World War.

This chapter explores one of the tropes of twentieth-century urban planning that envisioned an efficient, vertically separated network of mobility across the capital, as was seen in the previous chapter. But it does so from the civil engineer's point of view, rather than that of the urban planner, and situates it in some 30 years of concrete culture. This culture can be considered as a network of ideas, beliefs, and practitioners of civil engineering. It can also be characterized as a belief in the affordances of concrete to retrofit a new, efficient transport network into old urban environments. In considering this engineering culture of networks, this chapter offers a longer view of the ideas expressed repeatedly throughout the planning literature. It also provides evidence of the effects of London's distinctive political governance network on the realization of large and complex infrastructure projects, and the frustrations experienced when relying on public funding for traffic infrastructure. Finally, it examines the socio-political networks of modernity and democracy mediated by such structures, on both a global imperial scale and a local class-inflected scale.

Congestion at the Royal Victoria Dock, 1855–1915

In 1925, reporting to the Ministry of Transport on congestion in London's streets, its Chief Engineer, Henry Maybury, made the following stark assessment:

> The deplorable state of the roads-access to the Dock Area has long been a matter of great concern to the various authorities interested, and it is now generally recognised that nowhere in the whole of the London Area is traffic so congested or subject to such vexatious delays ... The worst congestion occurs on the way to Victoria Docks and there is nothing comparable to it in the whole of London.[2]

The vast Royal Group of Docks complex in Silvertown, London was constructed between the 1840s and the 1920s on previously deserted, low-lying marshland.[3] The North Woolwich Land Company, run by the engineer and property speculator George Bidder, purchased the area before building roads on the western and southern edges, along with the North Woolwich Railway (later part of the Great Eastern Railway), in 1847.

Along the riverfront numerous factories sprang up, housing noxious trades ousted from London proper by the 1844 Metropolitan Building Act. Charles Dickens described the area in 1857 as 'a place of refuge for offensive trade establishments turned out of town – those of oil-boilers, gut-spinners, varnish-makers, printer's ink-makers and the like'.[4] Henry Tomlinson, 1930s chronicler of the docklands, described Silvertown as:

> a perplexity of railway sidings, roads that wander and lose themselves, creeks intruding among backyards, cemeteries, gasometers, the funnels of steamers mixed with factory chimneys, warehouses built of the refuse of Tartarus, and a formless spread of the grey homes of those who, somehow, must supply the reason for its existence.[5]

Firms included the Thames Iron Works, Brunner Mond, Silver rubber works (hence Silvertown) and Henry Tate, as well as railway and telegraph works and a huge gasworks in nearby Beckton.

Bidder went on to promote the Victoria Dock, the first large enough for steamships, which opened in 1855. In 1864 it was sold to the London and St Katharine Dock Company (which owned dock complexes further upstream), which opened the Royal Albert Dock in 1880, at which time its predecessor was renamed the Royal Victoria Dock. In 1909 the newly formed Port of London Authority (PLA) took over all London's enclosed docks and built the King George V Dock, which opened in 1921. Throughout this development period, the rail network at the complex was expanded, with a huge marshalling yard eventually provided to the north-west of the site.

The Royals were far bigger than London's existing facilities, enabling larger and more numerous ships to load and unload their wares and, with excellent rail links and storage space, they soon specialized in goods-in-transit, primarily food. However, by the early twentieth century, whilst the complex was well served by rail, its road network had not been expanded to meet the demand brought by increased trade and the rise of motorized goods vehicles, and it had become one of the worst places in London for road traffic congestion. By the 1930s, the total volume of goods handled at the three docks exceeded 1 million tons per year, much of it transported by road, and was supplemented by rapidly growing traffic from the riverside factories.[6]

The Royal Docks lay in the County Borough of West Ham, just outside the LCC boundary, and the earliest coordinated demand for improvements to the access roads came in 1902, when a group of factory owners petitioned West Ham Corporation to look into the matter.[7] The formation of the PLA in 1909, and the rapid development of its plans to build the King George V dock, focused the merchants' attention on the problem. At a hearing of the committee that set up the PLA, the West Ham Corporation witness complained that 'we have been struggling about those level crossings for the last 20 years down there' and, in 1912, a group of some 25 firms led by the sugar giant Henry Tate and Sons (whose refinery was on Silvertown's riverfront) lobbied the Local Government Board, the Board of Trade, and the PLA to deal with the matter. Later that year, the West Ham Municipal Alliance submitted petitions containing the signatures of 1,870 local residents complaining of the delays, and a lively public meeting in the City of London brought many of the parties together to work out practical proposals. Over the ensuing months, the West Ham Borough Engineer and the PLA Chief Engineer worked together on an outline scheme to solve the problem with a new elevated road, which was put before the Road Board by a deputation on 15 July 1913, but was rejected, owing to the Board's existing commitments to building arterial roads.[8]

Interest in the scheme was growing but West Ham was having trouble stumping up its share of the cost. At a conference in November 1913 convened by the Local Government Board to look at London's road network, delegates heard about the proposal as an example of good ideas running aground in heavily taxed boroughs such as West Ham for lack of funds.[9] In 1914, the pressure group offered a class-inflected spin on the situation, remarking in a press release that 'Public feeling in the district affected particularly resents the attitude of the Road Board', which was 'favouring West London at the expense of East London' by funding arterial roads, 'which will be chiefly used for the purposes of pleasure' rather than urban main roads such as those in Silvertown, 'which are of far greater intrinsic value to the country'.[10]

At a subsequent Local Government Board spin-off conference looking at the specific needs of north-east London, the chairman remarked as follows:

it was the river that made London. London cannot exist without the river. The docks are an essential feature of both the river and London, and the growing population and the enormous accretion of trade and industry in the East and North-East of London call you who represent local authorities ... to take a big view of your duties and responsibilities with regard to the future roads and town-planning of this important area.[11]

The conference described the proposed Victoria Dock road improvement scheme as:

in any case ... a great blessing to the docks and to the factories at Silvertown, but in view of the additional dock accommodation to be provided, it now becomes of paramount importance, and should on no account be allowed to fall through.[12]

On 22 July 1914 a further deputation was made to the Road Board that this time resulted in an offer from the Board to contribute £100,000 towards the £300,000 cost.[13] The Board suggested that the remaining sum should be contributed by a consortium of the PLA, the West Ham Corporation, the Great Eastern Railway, and the manufacturers of Silvertown.[14]

Detailed discussions by a conference committee in 1915 crystallized into two distinct schemes, namely a relatively simple PLA-led proposal and a more ambitious West Ham Corporation plan, each involving new elevated roads to bypass the congestion black-spots.[15] The committee concluded that the simpler PLA scheme should be started immediately, with the West Ham proposals perhaps being added at a later date, although the Royal Institute of British Architects proposed a combination of both schemes that appears to have been ignored by the committee.[16] The PLA had stated that 'Victoria Dock Road is an evil of long-standing and one which requires drastic and immediate treatment.'[17] By this time, though, the First World War had intervened and costs for labour and materials began to shoot up, leaving the £300,000 scheme costs woefully short.

Developing a concrete proposal, 1919–1929

After the war ended, fresh attention was paid to the congestion in Silvertown. The old problems had not gone away, and with the PLA's King George V dock nearing completion, traffic began greatly to increase in the area. The 1919 report of the Road Board revived the 1915 arterial road conference proposals, and the establishment that year of a new Ministry of Transport galvanized action further.[18] The new Ministry's Director-General of Roads was Sir Henry Maybury, who had been Chief Engineer at the Road Board in 1913 when proposals for the Victoria Dock district had first been presented by the PLA, and he was instructed to prepare a detailed and costed scheme for improvements in the area, which he described in 1920 as containing 'congested roads carrying probably the heaviest traffic in the world'.[19] He also raised the scheme in his evidence to the Royal Commission on London Government, which reported in 1923, as one that demanded funding from across London as well as the Home Counties.[20] Later that year, as the London Chamber of Commerce was mobilizing the business community to carry out a major public lobbying campaign in support of road improvements at the docks, Maybury received full details of a workable scheme from his technical staff.[21]

Having first described the congestion in the area, their report went on to outline the proposed solution (shown in Figure 3.1).[22] Taking the problems in turn, the first proposal was to widen the eastern end of the East India Dock Road and construct a wide new bridge over the River Lea with easier gradients and straighter approaches than the existing bridge, known as the 'Iron Bridge'. Outline costings assumed this bridge would be made of steel, as was conventional at that time, but the report pressed for an alternative that was taking the engineering world by storm: reinforced concrete. Concrete was described in glowing terms, readers being informed that it 'would cost approximately two-thirds of the cost of a steel bridge. It would require less constructional depth allowing of better road grades while its upkeep would be nil and its life indefinite.'[23]

The next blockage, a narrow bridge over the Canning Town railway station, would be resolved by building a second new bridge with new station buildings set further back, enabling a wider roadway. The Barking Road east of Canning

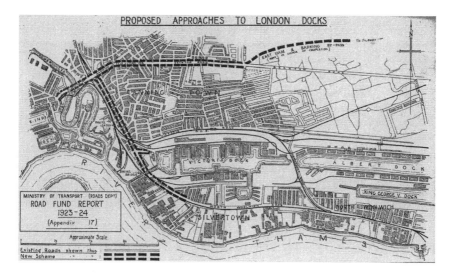

Figure 3.1 Ministry of Transport proposals for road improvement scheme in Silvertown, 1924, including bridge replacements, widenings, and an elevated flyover (Ministry of Transport).

Town railway station would be widened, with one consequence being that an easier turn south into Victoria Dock Road could be constructed.

But by far the most significant proposal was the construction of a wide elevated viaduct road along the line of the existing Victoria Dock Road, rising from ground level south of Barking Road, which would carry traffic over a critical level crossing known as the 'White Gates' as well as the entrance lock to the Victoria Dock (with its railway lines) before returning to ground level on the southern edge of the dock. It was to be an urban flyover. Spaces under the viaduct could be let to local industrial firms for parking their goods vehicles, while the existing streets at ground level could continue to be used by local traffic.

Maybury's scheme was well received, and was published in the annual report of the Road Fund, which by this time was administered by the Ministry of Transport.[24] The strategic nature of these traffic pinch-points was highlighted:

> The Iron Bridge ... is the funnel through which are poured all the streams of traffic bound for riverside and dock destinations from Poplar to Tilbury – sixteen miles as the crow flies, and including about 20 miles of river frontage.

At the White Gates level crossing, it was noted that 'the entire rail traffic to and from this dock area passes between these gates'. At this and other level crossings in the area, 'the confusion and congestion are almost indescribable'. All in all, the effects of these blockages became cumulative, as 'all these obstructions are so close together that when one is removed the flood of dammed-up traffic is immediately added to the block already forming against the next barrier'. And it was not just the delays to vehicles that were hampering trade in the area, but also delays to pedestrians, which were leading to bad timekeeping by factory employees. In summary, the Ministry argued, 'it can be said without exaggeration that delays here produce a retardation of movement through the entire metropolitan area'. But the cost had risen tenfold compared with the PLA's 1914 plan: the Ministry's 1923 scheme was estimated at 2–3 million pounds.[25]

However, by this time, the Ministry had another person in post with the influence to press for the scheme besides Maybury – namely Harry Gosling, who was the Minister of Transport, Whitechapel Labour MP, a former Thames lighterman, member of the PLA, and a prominent trade unionist who campaigned for better conditions for dock workers.[26] Now, having a detailed analysis of the traffic problem in the docks area, Maybury and Gosling together felt ready for action.

Two events occurred in 1924 that gave traction to the proposals. Firstly, Maybury presented a paper on the subject to the Royal Society of Arts (RSA), at a session chaired by Gosling, in which Maybury described the unchecked growth of the Silvertown area and the scant resources of the West Ham Corporation to provide for it, exclaiming that 'The result at Silvertown provides town-planners with conspicuous examples of what to avoid. For our ancestors' want of foresight the penalty has long been accumulating, and payment is now overdue.'[27] He described the 'disabilities the district now labours under, and what means of relief might be applied if London as a whole would put its shoulder to the wheel', before describing his scheme, which he believed to be 'one of the most urgent improvements conceivable'.[28]

In addressing the RSA, Maybury and Gosling had a specific goal in mind. They were presenting the scheme in order to seek financial contributions from across the capital to fund a road improvement that would be of benefit to all, not just the local area, with Maybury concluding that he was 'satisfied that the vital interests of the Imperial Capital demand the immediate inception of this scheme'.[29] He explained that his appeal was 'to London as a whole, London in its widest sense', with its 'incalculable resources'. It could not wholly be funded by West Ham, nor wholly by national government, but was 'first and foremost an All-London task, the cost of which should be distributed over a wider region than the Metropolitan Police Area'. In closing, he said that 'So great a campaign demands united forces led by a spirit of vision and venture. The Home Counties which derive so much of their wealth and rateable value from their vicinity to London, should not stand aloof.'[30] After a warm response from the audience, he finished with the comment that 'The country could not afford to have millions of money invested in those great Docks, and have the whole traffic from those docks throttled by the present wretched communications.'[31]

The second event in 1924 of significance to the problem was the passing, by Harry Gosling, of the London Traffic Act. This legislation established a London and Home Counties Traffic Advisory Committee (LHCTAC), bringing together government departments, London councils, the police, and a variety of local councils in nearby areas to offer a coordinated voice advising the Minister of Transport on traffic matters.[32] It had its origins in the 1923 Royal Commission on London Government, to which Henry Maybury had proposed the Victoria Dock Road scheme as an example of the need for a London-wide traffic body, and on its establishment he was appointed its first chairman. The LHCTAC was empowered by the Act to come up with proposals for sharing costs of road schemes among the various bodies and councils affected, and finally, here was an official (if only advisory) body with a mandate to seek London-wide funding, and one in which all interested parties could come together to coordinate their views. Within months of its constitution, the LHCTAC formally recommended the Victoria Dock Road scheme as a top priority.[33]

In 1926, the Royal Commission on Cross-River Traffic in London lent its weight to the pressure being exerted on government and local councils to fund the Silvertown improvement scheme, observing that 'The existing condition of affairs [in the Royal Docks area] has indeed become a public scandal.'[34] In 1927 a committee of MPs and parliamentary candidates for constituencies affected by the congestion pressed the government for action in response to the Royal Commission.[35] In 1928, with funding starting to come in from the various contributors, the government asked the LCC and the West Ham Corporation jointly to place a bill for the scheme before Parliament, which received royal assent in May 1929. A House of Commons select committee remarked that 'through the genius of the Minister of Transport, they had all the diverse authorities brought together and a practically agreed scheme of improvement'.[36] The £2.5 million cost was met by a grant from the government covering 75 per cent, with the remaining 25 per cent coming from the LCC; City of London Corporation; and nine nearby county boroughs and councils, including West Ham.[37]

The building of new roads, whether arterial or elevated, was inextricably bound up in this period with labour and unemployment policy.[38] Much of the road building programme of the 1920s and 1930s was paid for under unemployment relief policies, with the result that new roads did not necessarily map onto locations of congestion; or rather, *political* geographies of highway engineering did not always match *congestion* geographies. The former Road Fund secretary and roads campaigner Rees Jeffreys put it as follows in 1927:

> It must not be forgotten in reviewing the arterial road programme that it is a compromise between conflicting ideas and interests ... the Government, being concerned mainly with the relief of unemployment, were impelled towards projects affecting undeveloped lands free from those dilatory and costly obstacles which haunt all clearance schemes ... the undertakings which were ultimately selected were not always those holding first place on the traffic priority schedule.[39]

In later years, Jeffreys explained that this was why the Victoria Dock Road scheme took so long to realize. Unlike the arterial roads, which were built largely on virgin land, the dock road involved widespread clearance.[40] Congestion relief was not always a quick win or politically advantageous, especially in the shabbier parts of town, with their cramped spaces; working-class populations; and goods vehicles, trams, and buses jostling for space with motor cars. But the engineers kept the faith, and kept lobbying the politicians, even when circumstances changed. Henry Maybury retired from the Ministry of Transport in 1928, replaced as Chief Engineer by his protégé, Charles Bressey, but remained as the Ministry's consulting engineer and adviser for several years, as well as continuing as chairman of the LHCTAC, so the baton handover was a long one. After some two decades of promotion, Maybury was finally about to give Silvertown its flyover, with Bressey delivering the birth.

Building the Silvertown Way, 1929–1934

The Victoria Dock Road scheme delivered under Charles Bressey's leadership was divided into a series of packages, each put out to tender and managed by the scheme's consulting engineers, the firm of Rendel, Palmer, and Tritton, to which the Borough Engineer of West Ham, Lionel Jenkins, was attached.[41] First was the contract to build new housing for residents displaced by demolition. Between 1929 and 1931, 599 new cottages and flats were constructed in the nearby Prince

Regent Lane area on former allotment land largely donated by the PLA. The new estate was 'done on town planning lines', with tree-lined streets, a garden for each dwelling, and a ban on through traffic, giving it 'the appearance of a garden village'.[42] Next, the new bridge over the River Lea was constructed, capable of taking six lines of traffic, followed by a complex reconfiguration of the Canning Town railway station and its bridge. This work, begun in January 1930, was completed in September 1933. The biggest contract was the construction of the three-quarter-mile-long concrete flyover. Work on this began in November 1932 and was completed in 1934. Works elsewhere in the area formed a final part of the scheme. A total of 18 contracts were let, ranging in scale from £6,158 for preliminary street works to the £330,352 package for building the viaduct. Altogether 6,000 reinforced-concrete piles supported the scheme's foundations and each new structure was designed to carry a 100-ton vehicle on four wheels.[43] Figure 3.2 shows the newly completed flyover.[44]

Figure 3.2 Silvertown Way viaduct from the south-east, 1934 (Rendel, Palmer, and Tritton).

Silvertown Way was novel in two ways. Firstly, it was an early British example of a grade-separated urban flyover, an elevated road passing over not just a single river, road, canal, or railway (as in a bridge or simple viaduct) but lifting traffic over an entire ground-level transport network in a crowded city location. Secondly, it was realized largely in reinforced concrete, a technology that was reshaping civil engineering and architectural practice. The first reinforced concrete building in Britain, the 1897 Weaver's Mill in Swansea, was followed by a wide variety of projects at such a pace that, by 1926, the industry's journal was able to produce a vast conspectus of progress, covering buildings, bridges, viaducts, sports facilities, dock structures, industrial plants, highways, utilities structures, and sculpture.[45] Concrete was a modern material, and its industry, led by specialized architects, engineers, and contractors, worked with the energy and reforming zeal of pioneers. Its introduction and development also matched in date that of the motor vehicle, another symbol of dislocating modernity, and itself a technology that demanded new structural forms that reinforced concrete could create. Taken together, concrete and the motor vehicle created new urban architectures, with highway engineering emerging as a new discipline dedicated to reshaping the landscape of mobility.[46]

Silvertown Way was not the first urban concrete viaduct, but it was the most comprehensive and influential in Britain in the interwar period, being described by carscape historians Kathryn Morrison and John Minnis as 'the most significant step towards the post-war flyover' such as London's Westway, which opened in 1970.[47] Earlier examples included a viaduct in Dover, completed 1922; a system in Barking, completed 1927; and the Lea Valley Viaduct, also completed 1927, all of which bore structural and aesthetic resemblances to the Silvertown structures.[48]

In the United States, engineers had started building double-deck or viaducted urban streets such as Wacker Drive in Chicago, or streets with grade-separated interchanges such as the Bronx River Parkway in New York, in the 1920s (although the construction of concrete grade-separated freeways such as the Arroyo Seco in Los Angeles did not begin until the late 1930s).[49] Yet none of these early schemes matched the scale or complexity of the Silvertown scheme, which claimed to clear the most congested streets in the world and showed planners and engineers how vertically separated concrete roads in urban traffic black-spots could create a new network for traffic circulation.

Civil engineering historian Mike Chrimes has explained the role of early exemplars in propagating an engineering culture of elevated concrete structures, stating that 'As articles appeared describing the early system bridges, and textbooks became available, it became easier for local authority and other engineers to draw up their own designs', and indeed, by 1930, when the Silvertown Way was still under construction, some 2,000 reinforced concrete bridges had been built in Britain.[50] The Ministry of Transport had spearheaded this development, with one industry commentator observing in 1927 that some of its new bridges were 'of such importance that one can predict that reinforced concrete will be exclusively used in the future for bridges'.[51]

Silvertown Way provided a powerful boost to the engineering profession and its ambitions to reshape London, in its use of both reinforced concrete and its elevation. At a meeting of the Institution of Civil Engineers to discuss the scheme, one of Rendel, Palmer, and Tritton's engineers explained that concrete viaducts were chosen 'as they were felt to be the most attractive' in a scheme where 'continuity of design was of importance', and the structures 'were chosen to harmonize with their surroundings'. An important practical consideration in the Silvertown area, with its intensive industrial occupation, was that 'the air was chemically laden, which made the maintenance of steelwork difficult and expensive'. Another engineer singled out the high-level viaduct as 'an example which should be followed' by the Ministry of Transport, urging that further projects be developed employing the new approach – in particular that 'an attempt ought to be made to tackle one traffic-crossing right in the centre of London, no matter what the expense might be'. This, he hoped, would lead to widespread use of grade separation.[52] The concrete industry was actively promoting urban grade separation using concrete structures at that time, with one 1936 industry vision of a vertically segregated city (shown in Figure 3.3) influencing Alker Tripp's ideas on segregation introduced in Chapter 2, and considered in more detail in Chapter 4.[53]

With concrete, engineers were trained to conceive new forms of highway structures that reshaped the urban environment aesthetically and formally, such that, in the words of motorway historian Robert Baldwin, 'By the 1930s a professional understanding of the ... utility of concrete had become part of the highway engineers' professional discipline.'[54] Adrian Forty has gone further, arguing that reinforced concrete from the 1930s onwards created a new form of 'urban nature'.[55] Matthew Gandy has described 1920s American parkways as 'emblematic of a combination of engineering science with the aesthetic sophistication of landscape architecture' and 'a new spatial configuration of society, technology, and nature'.[56] It became natural, then, for engineers to create concrete landscapes in cities, as well as to consider concrete structures as viewing points for scenic vistas, a point to which we will return presently.

Highway Development Survey, 1934–1939

Following Bressey's completion of the Silvertown Way scheme in 1934, Leslie Hore-Belisha instructed him to carry out a comprehensive survey of highway developments needed in London over the following 30 years, as was discussed in Chapter 2. The *Highway Development Survey*, written with the help of architect Edwin Lutyens, was completed in 1937.[57] In Bressey's plans, we can discern a light-bulb switching on in the minds of highway engineers: that a new motor road network such as the Silvertown Way scheme could be

Figure 3.3 '"Upstairs footpaths" and shopping pavements, with cross-over bridges, visualized by the Concrete Association', 1936 (Concrete Association).

(literally) overlaid onto the existing London street pattern and, significantly, that this would be morally and aesthetically *good*. Bressey himself, discussing Silvertown Way with a group of traffic experts five years after it opened, said that:

> It affords a very instructive example of the advantages of a high-level road carried on a viaduct above the busy surface-streets teeming with local traffic. Incidentally it is to the credit of this viaduct that from it travellers gain a fascinating view, never before vouchsafed to them, of docks and shipping which are invisible from the streets at ground-level. Viaducts are so often and so thoughtlessly denounced as destructive of beauty that we

do well to remember the refreshing and hitherto unsuspected views they may reveal.[58]

At the same time, demonstrating the 1930s engineer talking like a 1940s or 1950s planner, Bressey described proposed new roads leading north from the dock area as follows:

> Many of the areas that would be traversed by these new routes are so woefully dejected and so stricken with decrepitude that the wholesale demolitions and consequent re-planning which the building of a wide new thoroughfare would entail could only be regarded as a boon that is long overdue.[59]

New roads cut through old London were, of course, nothing new; the novelty in the 1930s was the practical feasibility (demonstrated by the Silvertown Way) of new roads *flying over* the old, partly through the engineering affordances of concrete and partly through the handling characteristics of motor vehicles that could handle steeper gradients and tighter curves.[60] In a GPO Film Unit film made to publicize his report, Bressey lamented the failure of Christopher Wren to rebuild London on a rational grid system: 'a plan so enlightened, so far-seeing, that it would even have met the needs of modern traffic'.[61] Instead, the streets had remained unaltered – an opportunity lost – and the result, Bressey intoned, was congestion. 'Quick transport is essential to trade. Trade is London's life. Congestion strangles it', remarked the voice-over. 'But there is a way to ease this congestion. Just as Wren made a plan for seventeenth-century Londoners, so Sir Charles Bressey has worked out a plan for us.' Bressey is then seen looking down from a lofty office window onto the congested streets below, before placing a series of transparent overlay sheets onto a map of London, each sheet showing parts of his proposed new road network encircling and crossing the city.

Then Bressey described how his roads would cross each other and intersect with the existing street pattern in central London by using grade separation, gesturing to a series of dramatic visual depictions of aerial cityscapes. He said:

> We must build elevated highways on viaducts ... we must drive tunnels under obstacles ... we must have roundabouts and flyovers of the most modern type ... we must adopt every device for facilitating the movement of traffic and giving it the clearest possible course.

In his *Highway Development Survey*, and expressed vividly in the GPO film, Bressey was not so much proposing to *replace* the medieval street network of central London with a Wrenian grid of streets. Instead, he was proposing – as a highway engineer might – literally to *overlay* a new grid of concrete roads just above the old tangled surface of the city.

The moral, aesthetic, and material verticality present in Bressey's work had been generated too by a scientific gaze from above, in that he used aeroplanes to carry out aerial photographic surveys in an effort to prevent 'the delay necessarily attendant upon the ordinary survey methods'.[62] Vertical separation in London's streets was in the air in the 1930s, as Kathryn Morrison and John Minnis have observed.[63] It was an engineering solution, using concrete and steel, to the problem of London's ancient, complex, local street network. And, as a pragmatic engineering solution rather than an idealistic plan, Bressey's report urged the view that 'Londoners would be better advised to embark immediately upon useful schemes, admittedly imperfect, rather than wait for the emergence of some faultless ideal which will have ceased to be attainable long before it has received approval.'[64] The public, too, was thus being brought into a culture of concrete and the view from above. A concrete engineer reported in 1927 that 'The general public is gradually being educated into the many uses of reinforced concrete, and everyone is now well acquainted with this material, which is considered as a symbol of strength and durability.'[65] And this description – a symbol of strength and durability – could as equally apply to the British Empire, or rather the representation of it current in British culture at that time.

Democracy and 'a road to the Empire'

In 1935, the travel essayist Henry Tomlinson had narrated a fictional visit of a 'venturer' by train to the Silvertown docks, observing that:

> The miles of concrete quays, as he walks eastward to Gallions Station for a train back to town, will tell him more of Empire than all the exhibitions. For there are ports and ports. London is more than a seaport. It is a world market.[66]

With the opening of Silvertown Way, and the development of other concrete structures at the Royal Docks at the same time, this imperial concrete appeared more real than the concrete exhibition halls on the other side of the city in Wembley. The docks were the real heart of the Empire.

In July 1927, during a House of Commons debate on road funding, the trade unionist, West Ham Corporation member, and Labour MP for Silvertown, Jack Jones, had put on a piece of theatre for the amusement of his fellow members and the parliamentary reporters gathered in the Press Gallery:

> 'Not a mere arterial road', he said, swelling his breast and raising high his head, 'but an imperial road that Caesar would have gloried in'. He was half-serious, but he could not resist the temptation to talk in the towering way of the real Imperialists.[67]

In the months leading up to the completion of the Victoria Dock road scheme, West Ham Corporation and the Ministry of Transport had bickered about what it should be called. Early suggestions included 'King George V Avenue', 'Prince of

Wales Avenue', or simply 'London Dock Road'.[68] By 1 September 1934 the name had still not been decided on.[69] On 7 September, the Ministry's Deputy Chief Engineer, Frederick Cook, wrote to Leslie Hore-Belisha to say that he would be speaking to the West Ham Town Clerk the following morning to discuss the naming. He also told Hore-Belisha that 'Enquiries have been made from all sources in London concerning the Via-Dell'-Impero at Rome and a note is attached, giving the available information, together with a sketch illustrating the road.'[70]

The via dell' Impero, or 'Imperial Way', had been opened by Benito Mussolini in 1932 to reduce traffic congestion, but more importantly as a ceremonial avenue designed to bring into view Rome's ancient past. On 13 September 1934, Hore-Belisha proudly named *his* new road 'Silvertown Way'.[71] In 1926, as the Victoria Dock Road was debated in the House of Commons, Jack Jones had stated that 'it is not a road for West Ham, it is not even a road for London, it is a road to the Empire'.[72] The following year, as we have seen, he described the new scheme as 'not a mere arterial road, but an imperial road that Caesar would have gloried in'.[73]

Mussolini's Imperial Way, under construction at the time, was a road that connected the past to the future by creating engineered public views onto the products of Rome's ancient empire. Upon its opening, Mussolini had stated that:

> we must create the monumental Rome of the twentieth century. Rome must be a city worthy of its own glory. And this glory must be renewed incessantly so as to hand it down to future generations as the heritage of the Fascist Age.[74]

This creative destruction, a hallmark of urban modernity framed around modern problems such as congestion, was highlighted by the Italian archaeologist Guido Calza at the time, who noted that the via dell' Impero was not just an archaeological or aesthetic project, but 'a street necessitated by the traffic requirements of the modern city ... a modern street which also endeavours to give value to the remains of the Imperial Forums and monuments, to give a panoramic view of them'.[75] He concluded that, whilst it was a Roman road, it was 'no less an Italian road, the road of a modern nation, which joins modern initiative to the cult of her past'.[76]

The connection made by Jack Jones to Rome, and to empires, was picked up by Hore-Belisha, who, in his speech at Stratford Town Hall following the ribbon-cutting on the flyover, referred to Mussolini's Imperial Way as follows:

> Silvertown Way is surely as bold an undertaking as the Imperial Road of Rome ... It exposes to view not the ruins of the past, but a vision of our present maritime greatness ... This thing has been done by the people, by the so-called common men and women.[77]

Hore-Belisha made an important point about the public gaze:

> The London Docks are no exception to the general rule that docks and wharves are usually hidden behind high walls which shut them out of public view and are provided with fences ... The new road, with its bridges and its viaduct, will allow Londoners, and visitors to London, to gain a far-flung view of the waterways, docks and wharves which represent the real foundation of London's wealth. This is the Aladdin's Cave of London, and a thousand years of the romance of its commerce, wealth and industry are reflected in the scenes that are visible from these new viewpoints.[78]

It is hard to overstate the significance of this vision. As Felix Driver and David Gilbert have observed, 'The ebb and flow of goods at the docks determined the fortunes of hundreds of thousands of working-class people; there was in this sense no more significant site in the landscape of empire.' Many visitors to the area were incoming sailors or immigrant workers, with Driver and Gilbert noting that 'For many the first sight of London was of the vast and strange dockland landscape, on a ship sailing up the Thames estuary.'[79] But that first sight was not always a good one. Albert Linney, chronicler of London's river, described the old Victoria Dock Road as being 'as wretched a street as could be found in London ... What an impression of the Capital of the British Empire must have been given to foreign seamen emerging from the Royal Victoria Dock into London!' He went on: 'Every time I walked down that road to the White Gates, I was appalled at the squalor of the scene.'[80] Yet by this time, as well as hosting dock workers and transient seamen, this Aladdin's Cave also attracted tourists, with the PLA operating thrice-weekly summer steamboat excursions that included 'a run round the Royal Docks', attracting 1,000 passengers per day.[81] But these visions had always been at ground or water level. What could a view from above offer?

Adolf Hitler, too, had spent the early 1930s engaged in the construction of imperialist congestion-busting highway infrastructure, namely his motor roads, or *Reichsautobahnen*, programme initiated in September 1933. This project, as with Mussolini's Imperial Way, was held up for comparison with the Silvertown Way by those close to London's road building scene (although Hitler's roads were limited-access intercity highways rather than urban avenues). Rees Jeffreys, roads campaigner and secretary of the Road Board when Maybury was its Chief Engineer (and when the first proposals for the Victoria Dock scheme had been put forward), travelled widely in Europe and America examining their road infrastructure. Just hours after Silvertown Way was opened to the public, Jeffreys was onboard the German airship *Graf Zeppelin* along with several officials from the Ministry of Transport as part of the 1934 International Road Congress, hosted by Hitler and his deputy, Rudolf Hess. The engineers were inspecting progress on Hitler's road building scheme from above. Jeffreys was impressed by Hitler's single-minded ability to press forward his highway plans, building road capacity before congestion brought traffic to a halt rather than after, as Jeffreys complained had been the case in Silvertown.[82]

Mussolini and Hitler, as with the other dictators of interwar Europe, wanted to reshape their countries. But, as Eric Hobsbawm has pointed out, 'the antiquity of European civilization deprived them of the most obvious way of doing so: the building of entirely new capital cities'. So the solution, he said, came from engineers, not artists or architects: the building of new imperial *infrastructure*, whether it was the Moscow Metro or the Dneprostroi Dam.[83] To Hobsbawm's inventory we can add Hitler's *Autobahnen* and Mussolini's via dell' Impero.

Yet democrats wanted to create modern cities, too, and the Silvertown Way was an equally radical reshaping of old London: a modern engineering monument, in concrete, to the power of the British Empire. The fact that similar ideas about infrastructure could be manifest in markedly different contexts – places, politics, cultures – was partly a result of the networks of people and ideas that spread across Europe and America as examined by Driver and Gilbert, who note the profound influence of classical Rome on theorists of empire. They conclude that whilst 'There was no direct equivalent in London of, say, Mussolini's extravagant archaeological campaigns', influences from Rome could be found in London's sculpture and buildings.[84] Yet, to these artistic and architectural responses we can now add the engineering response manifest in the Silvertown Way, designed as a snub to il Duce's extravagance, an imperial way peculiar to London that opened up vistas not to an ancient past but to Britain's imperial present.

There was already, in fact, an Empire Way in London, part of the British Empire Exhibition of 1924–5 in Wembley – a complex expressed in reinforced concrete that offered visitors the chance to 'inspect the Empire from end to end'.[85] There was also, of course, the Mall, London's ceremonial avenue remodelled in the years after 1901 to commemorate Queen Victoria.[86] And one can readily see the London County Council's Kingsway, opened in 1905, as an imperial avenue.[87] All these imperial roads were consciously modern. Yet it was the Silvertown Way that we should see as an expression of a new imperialism – of free trade and openness gained through the unfettered circulation of people *and goods*, and by the opening up of the Empire to the public gaze.[88] This was a clear expression of modern democracy versus modern dictatorship. Driver and Gilbert have noted an anxiety that London's pre-eminence as the heart of empire might pass, like Rome's, owing to decay and neglect – which is exactly how the problem of congestion in the heart of the docklands was portrayed.[89] So it was Silvertown Way that was a riposte to Mussolini: a vast elevated reinforced concrete monument to a modern democratic empire; a new artery grafted onto the heart of empire to clear its congestion; new life in the face of decay, so as not to succumb to the ruination experienced in Rome.

And it was a Labour monument. On the matter of imperialism, Jones had set out the Labour view in his memoirs written in 1928. He characterized his party's policy as one based on ideas of self-government, especially in India, against the Tory 'govern and control' approach. Labour wanted to encourage an India along the lines of Canada and Australia – 'a free, independent, completely self-governing part of the Commonwealth of Nations which we call the Empire'. '*We ruled them for their good*', he parroted of the Tory view.

> It was the 'White Man's Burden'. Well, now, Labour says that if ever that was justified, it is so no longer, and that fashions in Imperialism, like girls' dresses, have considerably changed – and the whole idea is as considerably cut down!

Labour's imperial policy, he stated, was 'to trust the people on the spot, then; to develop self-government wherever possible; to repeal unfair treaties; to be truly democratic in foreign relationships'. Labour, he said:

> seeks earnestly to promote peace amongst all peoples and in co-operation to make the best of the resources of all lands. To do that she will abolish all artificial barriers, make intercourse free between peoples, in the firm belief that increase of knowledge of people and things is increase of understanding. She will endeavour to increase cheapness of transport and travel and abolish irritating restrictions.[90]

Conclusions

This chapter has examined the construction of a concrete flyover in east London that now attracts little attention but in its day was held up as a powerful statement of democracy at a time of rising autocracy in mainland Europe. Far from the British Empire being in decline in the 1920s and 1930s it was in fact a stronger cultural construct than it had been in the Victorian period. But it sat in a reformist discourse – reform from within – in which imperialism was fused with progress and modernity, and emphasis was placed on the machine and efficiency. The home city of empire was London, both modern and imperial, but it had to work well. There was no room for inefficiency. This was a working-class empire of free trade, communication, and technology: an empire of circulation for the improvement of all. This was an empire in which the goods circulating in Jack Jones's constituency of Silvertown represented a political stance, one in which the construction of better roads in modern materials such as concrete would reduce 'irritating restrictions' to world peace. And this was a circulatory empire peculiar to the 1920s and 1930s but one whose legacy would be felt after the Second World War, when the world had changed but ideas of urban circulation had become embedded.

For this imperial world view of efficiency, modernity, progress, and technology – a vision of free circulation – was also an engineer's vision. In an editorial published on the day the Silvertown Way was opened, *The Times* made a bold claim about the world view of the highway engineer, writing that 'those who see to-day the magnificent sweep of the new road will agree that the modern engineer is no whit inferior to the classical in his conception of what the beginning of an Imperial highway ought to be'.[91] It is clear how the ideas of Maybury and Bressey, with their reinforced concrete and ideas of new networks overlaid onto inefficient old ones, would take root in the political ferment of the interwar period, and flourish in the aftermath of the Second World War. This was a time of great social change too. The lofty view from the Silvertown viaduct over the docks, as it ran 'over the roofs and chimney-pots of Canning Town', enabled sailors to spot their ship at a glance, saving valuable time searching the vast dock complex, whilst lorry-drivers were able to make two more journeys per day between Silvertown and the City of London.[92]

The Victoria Dock Road scheme, with its elevated Silvertown Way, was unlike any other road building scheme in London in the interwar period. Unlike arterial roads, it was about freight rather than private transport; the working classes rather than the middle and upper (or rather, industry rather than leisure); a cramped, congested, complex urban streetscape rather than virgin agricultural lands; and messy local politics in a location (West Ham) that was not quite London but the heart of the Empire. As Jack Jones exclaimed, it was 'not a mere arterial road'. The docks were a part of London that few people saw but that affected everyone, and Silvertown Way afforded a new, touristic, *democratic* view into the Empire from above. And, as it soared over junctions and crossings, it was a distinctly *engineered* vision, a modern reinforced-concrete graft onto the Empire's congested old heart, a vision in which new technologies – concrete and the internal combustion engine – could enable the building of a new city of circulation.

Notes

1   *The Times*, 13 September 1934, 11; *Observer*, 16 September 1934, 16.
2   Ministry of Transport, *First Annual Report of the London and Home Counties Traffic Advisory Committee for the Year 1925* (London: HMSO, 1926), 32.
3   The history of the area is drawn from Bridget Cherry, Charles O'Brien, and Nikolaus Pevsner, *The Buildings of England. London 5: East* (London: Yale University Press, 2005), 290–303.
4   Charles Dickens, 'Londoners over the Border', *Household Words: A Weekly Journal* 16, no. 390 (12 September 1857): 241.
5   Henry Tomlinson, *Below London Bridge* (London: Cassell, 1934), 38.
6   Duncan Kennedy and Hubert Edward Aldington, 'Royal Docks Approaches Improvement, London', *Journal of the Institution of Civil Engineers* 2, no. 4 (February 1936): 6.
7   Donald McDougall, ed., *Fifty Years a Borough 1886–1936: The Story of West Ham* (London: West Ham County Borough Council, 1936), 175.
8   Museum of London Docklands Archive, PLA/CEN/1/2/4/11213, 'Delays at Swing Bridges and Level Crossings, Victoria & Albert Docks': correspondence, newspaper clippings, and extracts from minutes of evidence taken before the Joint Select Committee on the Port of London Bill, 23 July 1908.
9   Anon., *Report of the Conference on Arterial Road Communication in Greater London, Held at Caxton Hall, Westminster, on 25 November 1913* (London: HMSO, 1913), 20; The National Archives (hereafter TNA), MT 38/25: memorandum on RIBA report of PLA suggestions, 22 March 1915.
10 Museum of London Docklands Archive, 'Delays at Swing Bridges'.

11 Anon., *Report of the Sectional Conferences on Arterial Roads in Greater London, Held at the Offices of the Local Government Board on 9, 10, 11, 16, 17 and 18 March 1914* (London: HMSO, 1914), 7.

12 Anon., *Report of the Sectional Conferences on Arterial Roads*, 19.

13 Royal Commission on Cross-River Traffic, *Report of the Royal Commission on Cross-River Traffic in London: Minutes of Evidence and Appendices* (London: HMSO, 1926), 146.

14 TNA, MT 38/25, letter, PLA Chief Engineer to Road Board, 10 August 1918. The proposed figures were £75,000 from the PLA, £50,000 each from the Silvertown manufacturers and the West Ham Corporation, and £25,000 from the Great Eastern Railway.

15 Anon., *Report of the Sectional Conferences (Third Series) on Arterial Roads in Greater London, Held at the Offices of the Local Government Board on 19 and 28 May, 4, 18, 29 and 30 June, 26 and 27 July, and at the Town Hall, Bethnal Green, on 6 and 24 August 1915* (London: HMSO, 1915), 50.

16 Anon. *Report of the Sectional Conferences (Third Series) on Arterial Roads*, 54.

17 TNA, MT 38/25, letter, PLA to Local Government Board, 25 January 1915.

18 Road Board, *Ninth Annual Report of the Road Board* (London: HMSO, 1919), 26–59.

19 Royal Commission on Cross-River Traffic, *Report of the Royal Commission on Cross-River Traffic: Minutes*, 146; *The Times*, 20 April 1920, 11.

20 Royal Commission on Cross-River Traffic, *Report of the Royal Commission on Cross-River Traffic: Minutes*, 146.

21 *The Times*, 31 August 1923, 14; 1 September 1923, 5; 28 September 1923, 8; 24 October 1923, 11; TNA, MT 39/151, minute, Henry Tudsbery (Director of Engineering) to Henry Maybury (Director General of Roads), 11 December 1923.

22 Ministry of Transport, *Report on the Administration of the Road Fund for the Year 1923–1924* (London: HMSO, 1924), Appendix 17.

23 *The Times*, 31 August 1923, 14; 1 September 1923, 5; 28 September 1923, 8; 24 October 1923, 11; TNA, MT 39/151, minute, Henry Tudsbery (Director of Engineering) to Henry Maybury (Director General of Roads), 11 December 1923.

24 Ministry of Transport, *Report on the Administration of the Road Fund*, 22–5.

25 Ministry of Transport, *Report on the Administration of the Road Fund*, 24–5.

26 W. S. Sanders and Marc Brodie, 'Gosling, Harry (1861–1930)', *Oxford Dictionary of National Biography*, 2004.

27 Henry Maybury, 'The Victoria Dock District and Its Roads', *Journal of the Royal Society of Arts* 72, no. 3732 (30 May 1924): 464.

28 Maybury, 471.

29 Maybury, 468.

30 Maybury, 471.

31 Maybury, 474.

32 Great Britain, *London Traffic Act 1924 (14&15 George V Chapter 34)* (London: HMSO, 1924).

33 Ministry of Transport, *First Annual Report*, 5.

34 Royal Commission on Cross-River Traffic, *Report of the Royal Commission on Cross-River Traffic in London* (London: HMSO, 1926), 62–3.

35 *The Times*, 25 February 1927, 7.

36 *The Times*, 25 July 1928, 13; Great Britain, *Royal Victoria and Other Docks Approaches (Improvement) Act 1929 (19&20 George V Chapter 47)* (London: HMSO, 1929); *The Times*, 8 March 1929, 11.

37 Kennedy and Aldington, 'Royal Docks Approaches Improvement, London' (February 1936), 9–10.

38 See for instance Michael Hebbert, *London: More by Fortune than Design* (Chichester: John Wiley & Sons, 1998), 54; William Plowden, *The Motor Car*

and Politics in Britain 1896–1970 (London: Bodley Head, 1971), Part II; Andrew McDonald, 'The Geddes Committee and the Formulation of Public Expenditure Policy, 1921–1922', *Historical Journal* 32, no. 2 (1989): 643–74; G. C. Peden, *The Treasury and British Public Policy, 1906–1959* (Oxford: Oxford University Press, 2000), Chapters 4–5; Martin Pugh, *State and Society: A Social and Political History of Britain 1870–1997*, 2nd edn (London: Arnold, 1999), Chapter 11.

39 *The Times*, 28 October 1927, 15, in Rees Jeffreys, *The King's Highway: An Historical and Autobiographical Record of the Developments of the Past Sixty Years* (London: Batchworth Press, 1949), 208–9.

40 On the history of London local government, and its implications for road building, see John Davis, *Reforming London: The London Government Problem 1855–1900* (Oxford: Oxford University Press, 1988); Andrew Saint, ed., *Politics and the People of London: The London County Council 1889–1965* (London: Hambledon Press, 1989); and Hebbert, Chapter 3.

41 TNA, MT 39/389, 'County Borough of West Ham, Opening of New Road and Viaduct: Tuesday 13th September, 1934'.

42 *The Times*, 17 August 1929, 7; 16 May 1933, 11.

43 For extensive detail of the engineering of the Victoria Dock Road scheme, see Kennedy and Aldington, 'Royal Docks Approaches Improvement, London' (February 1936); anon., 'Bridge and Viaduct on the Royal Victoria Dock Approach Road', *Concrete and Constructional Engineering* 28, no. 5 (May 1933): 293–7; anon., 'Royal Victoria Dock Approach Road', *Concrete and Constructional Engineering* 29, no. 1 (January 1934): 77–80; and Duncan Kennedy and Hubert Edward Aldington, 'Royal Docks Approaches Improvement, London', *Surveyor and Municipal and County Engineer* 89, no. 2295 (17 January 1936): 79–81.

44 McDougall, 177.

45 Anon., 'Twenty-First Anniversary Number', *Concrete and Constructional Engineering* 21, no. 1 (January 1926).

46 On architectures of motoring, see Kathryn A. Morrison and John Minnis, *Carscapes: The Motor Car, Architecture and Landscape in England* (New Haven: Yale University Press, 2012) esp. Chapters 9 and 12.

47 Morrison and Minnis, 341.

48 Anon., 'New Reinforced Concrete Viaduct and Bridges at Dover', *Concrete and Constructional Engineering* 17, no. 6 (June 1922): 399–406; Mike Chrimes, 'The Development of Concrete Bridges in the British Isles prior to 1940', *Proceedings of the Institution of Civil Engineers: Structures and Buildings* 116 (November 1996): 408; Gordon Welch, 'Viaduct and Bridge at Barking', *Concrete and Constructional Engineering* 22, no. 4 (April 1927): 243–51; Public Works, Roads and Transport Congress, *British Bridges: An Illustrated Technical and Historical Record* (London: Public Works, Roads and Transport Congress, 1933), 106–7; Morrison and Minnis, 247.

49 Miller McClintock, *Street Traffic Control* (New York: McGraw-Hill, 1925), 55–7; Mark S. Foster, *From Streetcar to Superhighway: American City Planners and Urban Transportation, 1900–1940* (Philadelphia: Temple University Press, 1981), Chapter 5; Bruce E. Seely, *Building the American Highway System: Engineers as Policy Makers* (Philadelphia: Temple University Press, 1987), 150–6; Matthew Gandy, *Concrete and Clay: Reworking Nature in New York City* (Cambridge, MA: MIT Press, 2002), Chapter 3; Peter Norton, *Fighting Traffic: The Dawn of the Motor Age in the American City* (Cambridge, MA: MIT Press, 2008), 238–42; Thomas Zeller, 'Staging the Driving Experience: Parkways in Germany and the United States', in *Routes, Roads and Landscapes*, ed. Mari Hvattum, Brita Brenna, Beate Elvebakk, and Janike Kampevold Larsen (Farnham: Ashgate, 2011), 125–38; Joseph DiMento and Cliff

Ellis, *Changing Lanes: Visions and Histories of Urban Freeways* (Cambridge, MA: MIT Press, 2013), Chapters 1–2.

50 Chrimes, 411–12.

51 G. C. Workman, in anon., 'Twenty-First Anniversary Number', 101.

52 Kennedy and Aldington, 'Royal Docks Approaches Improvement, London' (February 1936), 35–6, 42.

53 *The Autocar*, 27 November 1936, 1107.

54 Sir Peter Baldwin, Robert Baldwin, and Dewi Ieuan Evans, eds., *The Motorway Achievement: Building the Network in Southern and Eastern England* (Chichester: Phillimore for the Motorway Archive Trust, 2007), 12.

55 Adrian Forty, *Concrete and Culture: A Material History* (London: Reaktion, 2012), 61.

56 Gandy, 122.

57 Charles Bressey and Edwin Lutyens, *Ministry of Transport Highway Development Survey 1937 (Greater London)* (London: HMSO, 1938).

58 Charles Bressey, 'Greater London Highway Development Survey: Discussion', *Geographical Journal* 94, no. 5 (November 1939): 356.

59 Bressey, 359.

60 On early London road improvement schemes, see for instance David Rooney, 'Visualization, Decentralization and Metropolitan Improvement: "Light-and-Air" and London County Council Photographs, 1899–1908', *Urban History* 40, no. 3 (August 2013): 462–82.

61 GPO Film Unit, *The City: A Film Talk by Sir Charles Bressey* (film), 1939.

62 On the role of aerial photography in London planning, especially post-Second World War, see Tanis Hinchcliffe, 'Aerial Photography and the Postwar Urban Planner in London', *London Journal* 35, no. 3 (November 2010): 277–88.

63 Morrison and Minnis, 338–9.

64 Bressey and Lutyens, 62.

65 G. C. Workman, in anon., 'Twenty-First Anniversary Number', 99.

66 Henry Tomlinson, 'Down in Dockland', in *Wonderful London: Silver Jubilee Edition*, ed. Almey St John Adcock (London: Amalgamated Press, 1935), 55–61, quote on 61.

67 *Guardian*, 6 July 1927, 4.

68 TNA, MT 39/389, minute, Ministry of Transport Secretary to Minister, 2 March 1934.

69 TNA, MT 39/389, minute, Ministry of Transport, 1 September 1934.

70 TNA, MT 39/389, memorandum, Frederick Cook to Leslie Hore-Belisha, 7 September 1934.

71 Newham Archives and Local Studies Library, 'Council of the County Borough of West Ham, Committee Reports, 1933–4', Vol. 48B (May–October 1934), 20 September 1934, 2216. The ribbon-cutting ceremony was captured in British Pathé, *Another Speed-Up* (film), 1934.

72 *Hansard*, HC Deb 14 July 1926, Vol. 198, col. 542.

73 *Guardian*, 6 July 1927, 4.

74 Anon., 'The via Dell' Impero and the Imperial Fora', *The Builder*, 23 March 1934, 496.

75 Guido Calza, 'The via Dell' Impero and the Imperial Fora', *Journal of the Royal Institute of British Architects*, 24 March 1934, 503.

76 Calza, 505.

77 *The Times*, 14 September 1934, 9.

78 TNA, MT 39/389, 'Victoria Dock Road: Notes for the Minister's Speech' (n.d.).

79 Felix Driver and David Gilbert, 'Heart of Empire? Landscape, Space and Performance in Imperial London', *Environment and Planning D: Society and Space* 16 (1998): 21.

80 *PLA Monthly*, November 1934, 6.

81 Adcock, 223.
82 *The Times*, 20 September 1934, 6.
83 Eric Hobsbawm, 'Foreword', in David Elliott, Dawn Ades, Tim Benton, and Iain Boyd Whyte, *Art and Power: Europe under the Dictators 1930–45* (London: Hayward Gallery, 1995), 15.
84 Driver and Gilbert, 'Heart of Empire?', 15.
85 Driver and Gilbert, 'Heart of Empire?', 23.
86 Tori Smith, '"A Grand Work of Noble Conception": The Victoria Memorial and Imperial London', in *Imperial Cities: Landscape, Display and Identity*, ed. Felix Driver and David Gilbert (Manchester: Manchester University Press, 1999), 21–39.
87 James Winter, *London's Teeming Streets, 1830–1914* (London: Routledge, 1993), 207–16; Dirk Schubert and Anthony Sutcliffe, 'The "Haussmannization" of London? The Planning and Construction of Kingsway–Aldwych, 1889–1935', *Planning Perspectives* 11, no. 2 (1996): 115–44.
88 On the wonder of aerial views on London, see David Gilbert, 'The Three Ages of Aerial Vision: London's Aerial Iconography from Wenceslaus Hollar to Google Earth', *London Journal* 35, no. 3 (November 2010): 289–99.
89 Felix Driver and David Gilbert, 'Imperial Cities: Overlapping Territories, Intertwined Histories', in Driver and Gilbert, *Imperial Cities*, 12–13.
90 Jack Jones, *My Lively Life* (London: John Long, 1928), 181–91.
91 *The Times*, 13 September 1934, 11.
92 *Manchester Guardian*, 14 September 1934, 8.

Bibliography

Museum of London Docklands: Port of London Authority Archive.

The National Archives: Ministry of Transport papers.

Newham Archives and Local Study Library: Council of the County Borough of West Ham Papers.

Hansard.

The Autocar.

Guardian.

Observer.

PLA Monthly.

The Times.

Adcock, Almey St John, ed. *Wonderful London: Silver Jubilee Edition*. London: Amalgamated Press, 1935.

Anon. 'Bridge and Viaduct on the Royal Victoria Dock Approach Road'. *Concrete and Constructional Engineering* 28, no. 5 (May 1933): 293–7.

———. 'New Reinforced Concrete Viaduct and Bridges at Dover'. *Concrete and Constructional Engineering* 17, no. 6 (June 1922): 399–406.

———. *Report of the Conference on Arterial Road Communication in Greater London, Held at Caxton Hall, Westminster, on 25 November 1913*. London: HMSO, 1913.

———. *Report of the Sectional Conferences on Arterial Roads in Greater London, Held at the Offices of the Local Government Board on 9, 10, 11, 16, 17 and 18 March 1914*. London: HMSO, 1914.

———. *Report of the Sectional Conferences (Third Series) on Arterial Roads in Greater London, Held at the Offices of the Local Government Board on 19 and 28 May, 4, 18, 29 and 30 June, 26 and 27 July, and at the Town Hall, Bethnal Green, on 6 and 24 August 1915*. London: HMSO, 1915.

————. 'Royal Victoria Dock Approach Road'. *Concrete and Constructional Engineering* 29, no. 1 (January 1934): 77–80.

————. 'The via Dell' Impero and the Imperial Fora'. *The Builder*, 23 March 1934, 496.

————. 'Twenty-First Anniversary Number'. *Concrete and Constructional Engineering* 21, no. 1 (January 1926).

Baldwin, Sir Peter, Robert Baldwin, and Dewi Ieuan Evans, eds. *The Motorway Achievement: Building the Network in Southern and Eastern England*. Chichester: Phillimore for the Motorway Archive Trust, 2007.

Bressey, Charles. 'Greater London Highway Development Survey: Discussion'. *Geographical Journal* 94, no. 5 (November 1939): 353–67.

Bressey, Charles, and Edwin Lutyens. *Ministry of Transport Highway Development Survey 1937 (Greater London)*. London: HMSO, 1938.

British Pathé. *Another Speed-Up* (film), 1934.

Calza, Guido. 'The via Dell' Impero and the Imperial Fora'. *Journal of the Royal Institute of British Architects*, 24 March 1934, 489–505.

Cherry, Bridget, Charles O'Brien, and Nikolaus Pevsner. *The Buildings of England. London 5: East*. London: Yale University Press, 2005.

Chrimes, Mike. 'The Development of Concrete Bridges in the British Isles prior to 1940'. *Proceedings of the Institution of Civil Engineers: Structures and Buildings* 116 (November 1996): 404–31.

Davis, John. *Reforming London: The London Government Problem 1855–1900*. Oxford: Oxford University Press, 1988.

Dickens, Charles. 'Londoners over the Border'. *Household Words: A Weekly Journal* 16, no. 390 (12 September 1857): 241–4.

DiMento, Joseph, and Cliff Ellis. *Changing Lanes: Visions and Histories of Urban Freeways*. Cambridge, MA: MIT Press, 2013.

Driver, Felix, and David Gilbert. 'Heart of Empire? Landscape, Space and Performance in Imperial London'. *Environment and Planning D: Society and Space* 16 (1998): 11–28.

————. 'Imperial Cities: Overlapping Territories, Intertwined Histories'. In *Imperial Cities: Landscape, Display and Identity*, ed. Felix Driver and David Gilbert, 1–17. Manchester: Manchester University Press, 1999.

Forty, Adrian. *Concrete and Culture: A Material History*. London: Reaktion, 2012.

Foster, Mark S. *From Streetcar to Superhighway: American City Planners and Urban Transportation, 1900–1940*. Philadelphia: Temple University Press, 1981.

Gandy, Matthew. *Concrete and Clay: Reworking Nature in New York City*. Cambridge, MA: MIT Press, 2002.

Gilbert, David. 'The Three Ages of Aerial Vision: London's Aerial Iconography from Wenceslaus Hollar to Google Earth'. *London Journal* 35, no. 3 (November 2010): 289–99.

GPO Film Unit. *The City: A Film Talk by Sir Charles Bressey* (film), 1939.

Great Britain. *London Traffic Act 1924 (14&15 George V Chapter 34)*. London: HMSO, 1924.

————. *Royal Victoria and Other Docks Approaches (Improvement) Act 1929 (19&20 George V Chapter 47)*. London: HMSO, 1929.

Hebbert, Michael. *London: More by Fortune than Design*. Chichester: John Wiley & Sons, 1998.

Hinchcliffe, Tanis. 'Aerial Photography and the Postwar Urban Planner in London'. *London Journal* 35, no. 3 (November 2010): 277–88.

Hobsbawm, Eric. 'Foreword'. In David Elliott, Dawn Ades, Tim Benton, and Iain Boyd Whyte, *Art and Power: Europe under the Dictators 1930–45*, 11–15. London: Hayward Gallery, 1995.

Jeffreys, Rees. *The King's Highway: An Historical and Autobiographical Record of the Developments of the Past Sixty Years*. London: Batchworth Press, 1949.

Jones, Jack. *My Lively Life*. London: John Long, 1928.

Kennedy, Duncan, and Hubert Edward Aldington. 'Royal Docks Approaches Improvement, London'. *Surveyor and Municipal and County Engineer* 89, no. 2295 (17 January 1936): 79–81.

———. 'Royal Docks Approaches Improvement, London'. *Journal of the Institution of Civil Engineers* 2, no. 4 (February 1936): 4–48.

Maybury, Henry. 'The Victoria Dock District and Its Roads'. *Journal of the Royal Society of Arts* 72, no. 3732 (30 May 1924): 464–74.

McClintock, Miller. *Street Traffic Control*. New York: McGraw-Hill, 1925.

McDonald, Andrew. 'The Geddes Committee and the Formulation of Public Expenditure Policy, 1921–1922'. *Historical Journal* 32, no. 2 (1989): 643–74.

McDougall, Donald, ed. *Fifty Years a Borough 1886–1936: The Story of West Ham*. London: West Ham County Borough Council, 1936.

Ministry of Transport. *First Annual Report of the London and Home Counties Traffic Advisory Committee for the Year 1925*. London: HMSO, 1926.

———. *Report on the Administration of the Road Fund for the Year 1923–1924*. London: HMSO, 1924.

Morrison, Kathryn A., and John Minnis. *Carscapes: The Motor Car, Architecture and Landscape in England*. New Haven: Yale University Press, 2012.

Norton, Peter. *Fighting Traffic: The Dawn of the Motor Age in the American City*. Cambridge, MA: MIT Press, 2008.

Peden, G. C. *The Treasury and British Public Policy, 1906–1959*. Oxford: Oxford University Press, 2000.

Plowden, William. *The Motor Car and Politics in Britain 1896–1970*. London: Bodley Head, 1971.

Public Works, Roads and Transport Congress. *British Bridges: An Illustrated Technical and Historical Record*. London: Public Works, Roads and Transport Congress, 1933.

Pugh, Martin. *State and Society: A Social and Political History of Britain 1870–1997*, 2nd edn. London: Arnold, 1999.

Road Board. *Ninth Annual Report of the Road Board*. London: HMSO, 1919.

Rooney, David. 'Visualization, Decentralization and Metropolitan Improvement: "Light-and-Air" and London County Council Photographs, 1899–1908'. *Urban History* 40, no. 3 (August 2013): 462–82.

Royal Commission on Cross-River Traffic. *Report of the Royal Commission on Cross-River Traffic in London*. London: HMSO, 1926.

———. *Report of the Royal Commission on Cross-River Traffic in London: Minutes of Evidence and Appendices*. London: HMSO, 1926.

Saint, Andrew, ed. *Politics and the People of London: The London County Council 1889–1965*. London: Hambledon Press, 1989.

Sanders, W. S., and Marc Brodie. 'Gosling, Harry (1861–1930)'. *Oxford Dictionary of National Biography*, 2004.

Schubert, Dirk, and Anthony Sutcliffe. 'The "Haussmannization" of London? The Planning and Construction of Kingsway–Aldwych, 1889–1935'. *Planning Perspectives* 11, no. 2 (1996): 115–44.

Seely, Bruce E. *Building the American Highway System: Engineers as Policy Makers.* Philadelphia: Temple University Press, 1987.

Smith, Tori. "'A Grand Work of Noble Conception": The Victoria Memorial and Imperial London'. In *Imperial Cities: Landscape, Display and Identity*, ed. Felix Driver and David Gilbert, 21–39. Manchester: Manchester University Press, 1999.

Tomlinson, Henry. *Below London Bridge*. London: Cassell, 1934.

———. 'Down in Dockland'. In *Wonderful London: Silver Jubilee Edition*, ed. Almey St John Adcock, 55–61. London: Amalgamated Press, 1935.

Welch, Gordon. 'Viaduct and Bridge at Barking'. *Concrete and Constructional Engineering* 22, no. 4 (April 1927): 243–51.

Winter, James. *London's Teeming Streets, 1830–1914*. London: Routledge, 1993.

Zeller, Thomas. 'Staging the Driving Experience: Parkways in Germany and the United States'. In *Routes, Roads and Landscapes*, ed. Mari Hvattum, Brita Brenna, Beate Elvebakk, and Janike Kampevold Larsen, 125–38. Farnham: Ashgate, 2011.

4   Cops, guard rails, and segregation[1]

Introduction

On Thursday 28 May 1936, the UK's Minister of Transport, Leslie Hore-Belisha, fixed into place a steel guard rail alongside the East India Dock Road, a busy traffic thoroughfare in Poplar, east London.[2] It was part of a 1.5-mile installation constructed that year at the request of the Metropolitan Police Traffic Commissioner, Alker Tripp, and its purpose was to prevent pedestrians from stepping into the roadway except at fixed pedestrian crossings. The barriers were just over 3 feet high and installed 12 inches from the edge of the kerb, enabling vehicles to drive close to the edge of the carriageway without fear of striking pedestrians, thus, it was said, making better use of road space. Gaps were provided at bus and tram stops, side roads and garages, and businesses requiring goods loading were provided with hinged sections under their lock and key for use during deliveries.[3] The entire installation was completed by September 1936.[4] It was the country's first ever large-scale installation of guard-rail technology (see Figure 4.1).[5]

Yet the guard rails were only the start of Tripp's plans. Soon after his appointment as Traffic Commissioner, Tripp had begun developing an ambitious London-wide scheme to save lives while keeping the traffic flowing.[6] It was to come in three phases. Firstly, he proposed building pedestrian crossings at fixed intervals of 150 or 200 yards along every main road in London. Secondly, he would erect guard rails along the entire length of each road, as at East India Dock Road, to prevent pedestrians from leaving the pavement other than at crossings. Thirdly, he would protect each crossing with a set of traffic lights (and, ultimately, barriers, as at railway level crossings). His physical technologies were then to be backed up by legislation, in that he wanted laws preventing pedestrians from entering the roadway except at the crossings.

Tripp's pedestrian control scheme was thus a profound departure from the laissez-faire traffic policing culture of his predecessors. It involved the physical partition of the streets and a radical set of prohibitions. Tripp believed that 'promiscuous pedestrian crossing will ultimately have to be prohibited',[7] and the word he consistently used to describe his approach was 'segregation'. By considering the wider context of this term, Tripp's pedestrian segregation can be seen to demonstrate links to racial segregation in the USA at this time, with a

Figure 4.1 Guard rails and pedestrian crossing on East India Dock Road, 1936 (William Whiffin, courtesy of Tower Hamlets Local History Library and Archives).

culture of categorization and exclusion common to both situations. Thus, in this chapter, we will explore how pedestrians on pavements were, in Tripp's intolerant and technocratic characterization, racialized, gendered, sexualized, aged, and classed. These pedestrians formed a diverse, populous east-end group whose supposed uncontrolled and disordered presence in the roadway was slowing those 'driving' through the streets, particularly relatively well-to-do – and therefore largely white and male – car drivers, whose vehicles travelled most quickly. Thus, in Tripp's view of automobilizing London, Poplar's pavement users were breaking down the 'proper' order of the road.

Tripp joined the Metropolitan Police as a Home Office civil servant in 1902. In 1920 he became chairman of the police recruiting board and in 1932 was promoted to become Assistant Commissioner with responsibility for traffic, a post he held until his retirement in 1947 (he died in 1954). He is often cited in the urban planning literature as the inspiration for ideas such as neighbourhood precincts, limited-access motor roads, and vertical separation. He appears in that literature as a marginal figure in the network of people and ideas, not quite a planner but strongly influential as a traffic practitioner. But, in a different literature – that of policing – Tripp occupies a more central position. Thus, so that we can understand his approach, Tripp's role as an actor in the culture of policing will be examined here, including the ways his police experience influenced his views on the pedestrian's place on London's roads, both from a safety point of view, and owing to more complex ideas developed through his experiences of American policing at a time of particular social and racial tension. This widening of the network, and the consequent centring of a marginal figure, draws a broader assemblage of concepts and beliefs into the story of London's traffic, helping us interrogate concepts such as pedestrian segregation that can easily be taken for granted.

Traffic control in the 1920s and 1930s

Alker Tripp's segregation proposal drew together into a London-wide network elements that were being introduced individually by the Ministry of Transport under the energetic leadership of Leslie Hore-Belisha. Trials of pedestrian crossings had begun in 1926 with large numbers installed from 1934. Guard rails began to be erected at busy junctions in central London in early 1935.[8] And Hore-Belisha's department had been erecting hundreds of traffic lights in London since 1932, following an experiment on London's Oxford Street in 1931 and a 1926 trial in Piccadilly in which the lights were provided to advise the police on point duty.[9] The novelty of Tripp's scheme was in its scale and in the systematic nature of his ideas – nothing like as large a scheme had ever been enacted in Britain, and his plan required rails, crossings, and lights to be interlinked and working together with pedestrians and vehicles as a single, techno-social system of segregation. But the problem was that the Metropolitan Police did not have direct control over traffic control installations. Responsibility lay with the patchwork of local authorities that covered London, and their design and specification lay with the Ministry of Transport. This was another case where London's governance frustrated attempts to take control – which was at the heart of Tripp's world view.

Traffic light systems in the 1930s could be of two main types: demand-based or time-based. The Ministry of Transport believed in demand-based traffic light systems. That is, it installed vehicle-actuated signals at busy junctions, with pedestrian crossings included in the junction layout and operated on demand by push-buttons. However, this technology was not without flaws. As the system responded to demand, rather than being on fixed time cycles, this meant that both drivers and those on foot faced unpredictable delays at light-controlled junctions, and sometimes these waits could be lengthy. According to the police, this led to pedestrians taking risks by crossing against the traffic, and drivers choosing to use side streets to avoid the hold-ups.[10] And, being a system relying on vehicle sensors, push-buttons, and interconnections between lights, the Ministry's vehicle-actuated technology was expensive. A Pathé newsreel film from 1933 depicts such an installation being ceremonially switched on, and the clicking of relay switches in the control box as vehicles drive over pneumatic sensors in the junction can be clearly heard.[11] This was a complex infrastructure.

Alker Tripp was outspoken in his criticism of the Ministry's use of vehicle-actuated signals. In 1935 he wrote to the Home Office as follows:

> The Ministry of Transport have prepared a scheme which to me appears to be a mere hotch-potch. They simply scatter traffic lights all over the map at any likely looking crossing. I think this is quite useless, as numerous unrelated traffic lights impose so many jerky stoppages on the traffic that drivers go round the back streets in order to avoid the main routes and this creates new dangers.[12]

He had previously written to a Ministry of Transport official, as follows:

> the whole picture looks to me to be not a cohesive system, but a widely scattered sprinkling of independent installations, which might indeed ultimately prove mischievous rather than helpful. What we want to secure is regular passage for pedestrians; and pedestrians must not be left guessing as to the intervals between the successive stoppages of the traffic or the length of time that will be given to them for passage. From that point of view vehicle-actuated signals are far from ideal in inner areas.[13]

Tripp's proposed scheme was different. As we have seen, it comprised a network of guard rails, pedestrian crossings, and traffic lights set up at fixed intervals *between* busy junctions, with signals on the flexible progressive system, which was time-based rather than demand-based. This took the form of a network of traffic lights, each timed to synchronize with others on a route to create a green wave of light phases progressing along the road at a fixed speed – say, 17 mph. A driver travelling at the system speed should therefore expect to find all lights at green as the journey progressed. In effect there would therefore be platoons of vehicles moving at a fairly constant speed, separated by regular gaps in the traffic coinciding with the red phases. The pedestrian phases at each crossing would be timed to occur during the gaps – reducing stoppages of traffic – and the gaps would also assist traffic turning into the main road from side streets. The result would be consistency of flow, predictability of waits for both vehicles and pedestrians, and the minimum of hold-up.[14] Tripp's traffic lights were also cheaper than those of the Ministry, which he hoped would enable them to be installed systematically across London, rather than being 'scattered' where local conditions dictated. Tripp saw London's traffic as a single system; he was a system-builder.[15] But by late 1935 he was getting nowhere with the Ministry of Transport on his London-wide scheme so, to break the political logjam, he proposed a 1.5-mile experiment on a single busy trunk road in east London, the East India Dock Road.

The East India Dock Road experiment, 1936

The East India Dock Road was chosen for the experiment as it had a high pedestrian accident record.[16] But the project was about more than road safety. It was a political struggle for control of London's streets, in two ways. Firstly, it was a struggle between the Metropolitan Police and the Ministry of Transport. This was politics played out at the highest levels. But Tripp's project was also a political struggle played out on the streets of east London, in its local council committee rooms, shops, businesses, chambers of commerce, and the pages of its local newspapers. Tripp wanted to seize control of London's streets from elected authorities, and he began his assault in Poplar. He had a fight on his hands, telling the Metropolitan Police Commissioner, Hugh Trenchard, 'if we are to have any real system for London, we cannot continue to be fettered by the whims and fancies of the numerous Local Authorities concerned'.[17]

But he did continue to be fettered. The Transport Minister, Leslie Hore-Belisha, had been receptive to Tripp's ideas for the East India Dock Road, but felt that the experimental system was too big to ask the local authorities to swallow in one go – and that it was too expensive for the Ministry to fund itself from the Road Fund, which would have had far-reaching political implications on spending outside the capital. The compromise was a split. The Ministry would pay a portion of the cost of erecting rails to represent the wider benefits to London, and the local authorities through which the road passed (Poplar and Stepney Borough Councils) would pay a share representing the local benefit to residents and businesses. Traffic lights, it was proposed, would be dealt with later once the rails were up. But Poplar and Stepney were not easily convinced, and various deals had to be offered and negotiated during autumn 1935 before the councils grudgingly accepted and work could begin.[18]

To explore the effect of the East India Dock Road guard rails, we may look at three main sources of commentary: the press, motor and pedestrian lobbies, and local business organizations. Much of the press reaction to the scheme took Tripp's line, which was that road traffic and foot traffic should be separated. This view assumed the road was a traffic artery. The *Manchester Guardian*, for instance, predicted that Britain was:

> probably approaching a time when no pedestrian will be allowed on the roadway at all (at least in towns) except at stated points under special protection. If motor transport continues to expand at its present rate the segregation of urban wheeled and foot traffic must some day be complete.[19]

Once the Poplar rails were up, the same newspaper suggested that 'the principle of putting something more solid than common sense between the motorist and the pedestrian is bound to grow in strength'. It went on to state that 'Whether one will consider it a division of the sheep from the goats or the lambs from the slaughterer will depend, no doubt, on which side of the rail one finds oneself.'[20] This might be seen as a view driven by the motor lobby, and the car interests certainly approved of guard rails as a means to shift blame for accidents from motorists to pedestrians.[21] The Automobile Association, for instance, lobbied for guard rails on the basis of their successful use in mainland Europe, which had a tendency 'to deal with the pedestrian as a traffic unit rather than a separate problem'.[22]

To find a countervailing view, one needs to look at two groupings who might have felt aggrieved at this interference in the conduct of the streets. The first is the pedestrian, and the response of the Pedestrians' Association (now Living Streets), founded by Viscount Cecil of Chelwood in 1929, vacillated between qualified support and trenchant criticism. The association had already expressed unease in 1934 about the idea of non-junction pedestrian crossings (known then as 'crossing lanes'), fearing that plans to make their use compulsory would be 'a grave interference with the liberty of the public', commenting that:

> a man whose office or shop is midway between two crossing lanes would have to walk a quarter of a mile to make a call on the opposite side of the road and to return, or a householder would have to walk the same distance to post a letter in a pillar box on the other side of the road.

The association mobilized age, class, and gender categories as follows:

> If the crossings are widely adopted, it is right that pedestrians, and particularly children and old people, should be urged to use them as much as possible, and they will have the strongest inducement, that of self-preservation, to do so – but it is unreasonable to ask busy business men to go considerable distances out of their way ... The prohibition would slow up pedestrian traffic enormously and cause congestion.[23]

The result of the association's representations to Hore-Belisha on legislative prohibition was positive: he abandoned the idea. But when, the following year, the East India Dock Road guard rail experiment was announced, the association swiftly denounced it as a material alternative to legislative prohibition, stating that:

> it represents a blow at the existing rights of foot passengers, and imposes on them an enormous inconvenience. It is one more step in the policy of handing over the roads to the motorist and depriving all other road users of their rights.[24]

Yet it went on to produce a report for local authorities agreeing that 'anything that deters persons, especially the young, from stepping thoughtlessly from the pavement on to the roadway will tend to render less frequent one common form of accident', so long as the goal was to promote safety, not speed. Its biggest concern was the danger to bus passengers boarding or alighting outside the guard rails, but it was also keen to stress that there must be 'no legal limitation on a pedestrian's right to cross the road at any point'. A further cause for concern was, perhaps surprisingly, the convenience of the road-user:

> Has he to stop opposite the house or shop at which he wishes to call and to climb through the railings, or to shout and gesticulate until he can attract the attention of the frontager to come with a key and remove the railings?

The association pressed for the rails to be removeable by road users every few yards, and advised an evaluation of the experiment after 12 months.[25]

By 1938, the association's position had hardened into a belief that the continuous guard rail idea was part of 'the movement, encouraged by the motoring organisations, for limiting the pedestrian's right to the use of the road'. It recognized that 'modern conditions call for a certain measure of regulation of the movements of all road users', including the use of short sections of railing at particularly dangerous junctions or crossings, but that 'the general invasion of pedestrians' rights to the use of the Highway under the guise of safety measures as an alternative to the restriction of speed is a policy which the Committee has stoutly resisted and will continue to resist'.[26] A year later, it concluded that the experiment was dead, with no further installations planned, as 'the rails proved very unpopular, for they not only caused great inconvenience, but in some respects added to the dangers'.[27]

A similarly equivocal reaction came from the second grouping for whom the street was a place as much as an artery, namely the local business community. The East India Dock Road was largely commercial.[28] For these people, the street was an intensely local place, control over which was not to be given up lightly. In 1936, as the rails were being erected, the Poplar Chamber of Commerce passed a resolution urging that the local councils consult with local businesses before starting work on the guard rail scheme, observing that 'customers, rather than climbing through the rails preferred to go further for their purchases instead of dealing with their regular tradesmen'.[29] Yet even this response was muted, and was more a chance to criticize the local councils than a sustained campaign against prohibition. It certainly was not a criticism directed at either the Metropolitan Police or the Ministry of Transport.

With opposition from local authorities over financial contributions, Tripp's final hope in getting the experiment completed rested in convincing the Ministry of Transport to shoulder the whole cost of his rails-and-lights experiment. He remained in no doubt that this would be the right thing to do, explaining in frustration the following:

> Some millions of money are being spent from the Road Fund to provide new roads, new bridges, and fresh avenues of circulation for motor traffic thus increasing the length of the cutting edge of the tool that is inflicting such a heavy toll of life and limb, but money is denied for a scheme of public safety.[30]

However, by the time the rails were up in the East India Dock Road, he was losing political support. Trenchard was replaced as Police Commissioner in late 1935 by the former Governor of New South Wales, Sir Philip Game, and Hore-Belisha soon left the Transport Ministry, replaced in 1937 by the lacklustre Leslie Burgin. Neither Game nor Burgin appeared to have the appetite to continue the road safety work of their predecessors with as much vigour, leaving Tripp to fight alone.

By 1938, responding to a request for an update on the traffic lights, Tripp wrote a plaintive note to Game: 'I am almost in despair about this matter.' He observed that Trenchard had compared London's slow progress in installing lights with the speed of progress in New York as long ago as 1932, noting that New York had a 'definite co-ordinated plan' whereas the 'few in London are on no general plan at all'. For Tripp, the scheme was 'aimed at pedestrian safety equally with (or even more than) vehicular circulation' but, since the guard rails had been erected, 'the effect has been to localise the casualties at the crossings. The protection of those crossings [with traffic lights] is obviously the next move.' He concluded that the police were 'powerless without the aid of the Ministry of Transport and the Local Authorities, and, if they show so little response, there is not much hope of really getting the casualties down'. The solution, Tripp had long argued, was for the police to take control of traffic across London away from local authorities, placing London 'in the same position as the Police in other large cities, such as New York, Chicago, Berlin, Paris, etc.'.[31] But it was not to be. The outbreak of the Second World War put all plans on hold, with the experiment only half built.

It was easy for Tripp to knock the local authorities, frustrated by having to negotiate with so many of them for a single scheme. Yet schemes that affected local people raised local issues that the councils were elected to consider, and in this episode, we clearly see the tension between framing roads as arteries in a closed system and seeing them as a succession of *places*: of localities, work-places, shopping areas, or neighbourhoods. This was a period in which the latter view was being 'persistently, if unevenly, revised', as historian Richard Hornsey has described, as 'the street became culturally redefined no longer as a site of heterogeneous social activity, but as a functional conduit whose purpose was to facilitate the speedy flow of traffic'.[32] An examination of the way London's streets were policed helps illuminate this shift.

Policing traffic in the motor age

When the motor car entered London's traffic scene at the turn of the twentieth century there was no government transport ministry or professional transport engineers, and the urban planning profession was, as we have seen, in its infancy. The obvious body to control traffic London-wide had long been the Metropolitan Police, but it had grown up in a culture of laissez-faire.[33] In the 1860s, when London's streets were becoming increasingly congested by commercial vehicles, buses, and cabs, Sir Richard Mayne, the joint first Commissioner of the Metropolitan Police, held that 'he did not think the wit of man could prevent by any amount of regulation the slowness arising where there is a great traffic along a narrow line; all that can be done is to prevent absolute obstruction'.[34] The general view of the police by 1900 was that its role was to remove blockages, separate different classes of vehicle according to their speed, and simply keep traffic moving.

James Winter has provided an authoritative account of the tension in London's policing scene before the motor age.[35] From the start, Winter observes, Robert Peel 'recognized that, in an open society where the desire for order and the desire for personal freedom must be in permanent tension, a police force must seek to find an equilibrium'.[36] Winter expresses the paradox as 'wanting policing to be both "mechanical" and "social"' while the police commissioners had 'no intention of trying to solve it'.[37] This was a liberal paradox and it underpinned London's traffic scene profoundly. Nowhere was this tension experienced as much as on London's streets, as personal freedom – the ability to move freely – was only possible if individuals were controlled by an outside force: the state, the police. Otherwise one person's freedom created another person's restriction – congestion. Winter reports a set of leader articles in *The Times* that lamented the irreconcilability of the paradox, that the 'exercise of an unrestricted right to come and go as one liked or dared might actually decrease the sum of individual liberty'.[38] He then summarizes the legislative instruments developed through the later nineteenth century to manage London's traffic, but all with a keen eye on ruling with the lightest touch. Throughout was an assumption that pedestrians had rights over vehicles to occupy the streets, and that a crucial role of the police was to protect the pedestrian from harm – albeit tempered by an unwillingness to submit drivers to undue restriction.[39] Carlos Lopez Galviz has further examined the Victorian traffic police officer's paradox, noting that 'The issue of civil liberties and the degree to which they could be, and were indeed, negotiated on the streets made the introduction of any form of regulation a challenging affair.'[40]

With this liberal tension as a backdrop, we move into the twentieth century, and the reaction of the Metropolitan Police was an amplification of the classic tension as vehicle speeds and handling characteristics began rapidly to change. The coming of the motor age became characterized in two key ways. Firstly, and most commonly, it was a matter of accidents, injury, and loss of life. In this characterization, speed – the main cause of accidents – was bad, and thus congestion was good: it slowed traffic and therefore prevented accidents. But the second characterization of the motor age was the old desire for free circulation, and here, congestion was bad, as it was circulation's 'inextricable other', as Galviz has put it.[41] We thus see that in the context of the police, the discourse of congestion was intertwined with that of speed, circulation, danger, and death.[42]

Two internal histories of the Metropolitan Police's traffic department have picked up the story of the twentieth century.[43] In them is described a fragmented development in the first two decades, as responsibilities for traffic management accreted into one department created in 1919. This department was designed to be in close contact with the government's Ministry of Transport, also created that year. Originally an administrative and advisory body, by 1923 the police traffic department started to take a more active role in governing London's streets, with Arthur Bassom, a traffic superintendent known for his knowledge of London traffic, appointed as the first Director of Traffic Services.[44] Major changes occurred the following year with the passing of the 1924 London Traffic Act, which created a more active role for the traffic department as well as the body to liaise between the police and the ministry known as the London and Home Counties Traffic Advisory Committee, on which Bassom sat, discussed in the previous chapter. For the first time, London had a capital-wide traffic body with Metropolitan Police traffic experts playing a key role, although its powers were diluted by the breadth of its membership.

Yet the rapid and widespread adoption of motor-cars after the First World War profoundly reshaped the relationship between the police and the public, as Clive Emsley has described. 'The development of motor vehicles travelling at much greater speeds than previous road traffic', he says, 'constituted a problem of a new dimension. By the early 1920s the use of the law to control motor vehicles was jamming the magistrates' courts and creating friction, hitherto unknown, between the police and the middle classes.'[45] William Plowden's account of the motor-car and politics further highlights changes in the public–police relationship, with increasing regulation in the 1920s and 1930s.[46] And Keith Laybourn and David Taylor have provided a detailed account of the effects of motorization on policing in the interwar period, describing governments as 'convulsed with the fear of communism and consumed with concern about motorised transport and the "road holocaust"'. They note that:

> the seemingly narrow issue of the regulation of traffic raised wider questions of individual freedom and, much to the concern of the police, brought ordinary, working-class constables into contact (and conflict) with members of the middle classes, who previously had held a positive, if somewhat patronising, perception of the British bobby.[47]

Alker Tripp was clear about the role the motor car had played in reshaping the relationship between the police and the public. 'Motor-car law', he said in 1928, 'has brought a new stratum of the public into frequent relations with the police. The area of contact, and of potential friction in consequence, increases daily.' He went on to note that:

> the individual constable enters on the task already handicapped by having to some extent lost his reputation as a monument of stolid and tolerant commonsense and a barrier against interference with individual liberty. He is too often regarded as a busybody, a stickler for trifles ... The police must vindicate anew their quality in the public eye.[48]

Tripp's pedestrian control scheme was thus a radical departure from the traffic policing culture of his predecessors. It involved the physical separation of the streets and a radical set of prohibitions. Tripp believed that 'promiscuous pedestrian crossing will ultimately have to be prohibited'.[49] Yet he favoured the employment of technologies that embodied that control in the public itself.

Miles Ogborn has analysed traffic lights in this respect, describing them as 'a branch of the undercover police' that 'can organise the chaos of urban individualism on a grand scale'. 'In all this', he says, 'the traffic-lights signal the dream of the liberal state. Their aim is to disappear. They work best when no one notices them, when regulation becomes co-ordination and facilitation.' Yet, as he observes, 'It is only under certain political conditions that the signals work.'[50] In this mode, then, traffic lights reshape the public as much as the relationship between the public and the police – they internalize the police in the public. More broadly, Christopher Otter has analysed the role of infrastructural technologies in embodying control and normalizing behaviour, stating that 'Technology was, and remains, a fundamental way of liberally governing the social.'[51] Patrick Joyce, too, has examined the relationships among technologies, governance, and liberalism, concluding that it was in 'the republic of the streets' where 'subtlety and ambiguity were most apparent'.[52] Carlos Lopez Galviz has further reinforced the idea.[53] And Thomas Osborne and Nikolas Rose have discussed the 'regularisation of liberty' in the context of 'fears of the mob, the mass, and riot', which suggests an intriguing wider context to ideas of motorist and pedestrian control in London in the 1920s and 1930s.[54]

What Tripp really wanted was a grid-plan city like those in the USA, which offered multiple routes to bypass congestion and, in their regularity and order, lent themselves more readily to control. By the 1930s, US city grids had come to make great impressions on visitors from places such as London.[55] Tripp could not, of course, reshape London's street plan literally. Instead, he proposed to replicate in the British capital's streets the moving platoons of vehicles and pedestrians that occurred in the gridded streets of New York, Chicago, and Detroit. This was consciously understood. A Ministry of Transport official said Tripp's pedestrian scheme was designed 'to split up the distance between signalling installations at important cross roads into a number of equal distances, so that, as in the block layout of streets in America, an ideal system of flexible progressive working will be possible'.[56] Predictability; a sense of systematic order; and centralized, absent control were key to Tripp's scheme, rather than the system merely responding to the actions of individual actors on the street, as the vehicle-actuated signals favoured by the Ministry of Transport appeared to do. By introducing American methods to London's streets, Tripp was attempting to smuggle in their police forces' city-wide traffic control – and he had seen what he wanted at first hand.

Prohibition and the police state

One of Tripp's first major tasks as traffic commissioner had been to go on a fact-finding visit to the USA and Canada to study control methods, which took place over 20 days in October 1934 and included visits to New York, Boston, Detroit, Chicago, Toronto, and Montreal. Upon his return to London, Tripp prepared a substantial report for his police colleagues and the Ministry of Transport detailing his findings. The tension between accidents and free-flowing traffic was expressed clearly in the report, which was split into two main parts – accidents and circulation. But the report conveyed a wider set of issues that shed light on the London traffic scene post-1934, which – as we have seen in Chapter 2 – Tripp significantly shaped.

For Tripp, the difference between America and England was the difference between prevention and cure. 'One is conscious', he began, 'of what appears to be a constant solicitude on the part of the House of Commons for the motorists' interest – a solicitude which arises largely from the desire to protect personal liberty'. He went on:

> the Americans seem in one important particular to have been clearer-sighted than ourselves: they have consistently relied upon prohibitions intended to prevent bad habits of driving (e.g. by speed limits, stop-streets, etc.) instead of merely leaving dangerous driving to be dealt with after it has occurred.[57]

Tripp called for immediate action:

> Such action lies along the lines of (i) physical safeguards – traffic lights, duplicate carriageways, railings ... and (ii) a code of exact law (based on the advice of traffic experts to the exclusion of interested parties) which is firmly, rigidly and exactly enforced by a more specialised police organisation with the help of courts which are more alive to the public danger and less concerned with the liberty of action of the motorist.[58]

To get this action, London required one traffic authority. Instead, following what Tripp saw as damaging legislation in the 1924 London Traffic Act, which could have created such a body, London had 'local traffic regulations dependent on our troublesome, dilatory and old-fashioned machinery of Local Government. There are 132 Local Authorities in London.'[59]

Tripp was agitated by what he saw as the lost opportunity of 1924. Traffic control on the streets was, he believed, 'inevitably a matter which must be decided (whether in America or England) on the word of the Police authorities who alone are responsible for day to day handling of traffic'. He continued:

> the responsible Police Authorities of a great City ... remain the only effective regulating agent. New York recognises this fact openly; London (under the Act of 1924) camouflages it under a system of circumlocution which delays, complicates and sometimes aborts – but is never constructive.

He concluded:

> True, there are many local interests to be balanced and these often conflict; but it is those very interests that the Police are already balancing, day in, day

out, when they deal with street parking, loading, obstruction and a score of other cognate matters.[60]

For Tripp, a crucial problem in London was the relationship between the Metropolitan Police (under the Home Office) and the Ministry of Transport. By 1934, Tripp was clear that:

> the functioning of the Ministry of Transport, a Ministry created to effect unification, has – so far as Traffic in London is concerned – effected partition instead. It has driven a line of cleavage between the executive and administrative sides of Traffic. The administrative cannot move hand or foot without the executive, and the latter is under another Government Department (the Home Office).[61]

Tripp, in London, saw 'over-complicated machinery' that produced both 'confusion' and 'paralysis', by contrast with America. 'Traffic policies must be widely conceived', he explained, 'and to have to argue and negotiate every smallest detail with a series of local bodies who have themselves no responsibility for traffic is indefensible'. He continued:

> The Police are now [since 1924] made to appear merely as one voice among many, and both the public and the Local Authorities will argue and go on arguing on equal terms with departments in cases where they would accept as law the verdict of the Commissioner of Police, knowing that the Police are not only the practical people, but are the only people responsible for getting the traffic through.[62]

He concluded that the machinery of the 1924 Act was 'a mischievous sham'.[63] This sets into context Tripp's frustration over the following two years with his pedestrian control scheme in the East India Dock Road. But there was more to his frustration than traffic schemes and recalcitrant local boroughs. He wanted to change London's culture of policing.

By 1932, when Tripp was appointed traffic commissioner, the world had changed dramatically compared with a generation before. The rise of the motor-car city had coincided with the arrival of the political labour movement and the development of an increasingly interventionist state. Martin Pugh has surveyed a major social and political shift at the start of the twentieth century, and describes what he terms 'a sustained extension of the role of the state between 1906 and 1914'.[64] Fears of racial deterioration following the Boer War (in part) led to demands for more state intervention in everyday life, and the demands of the First World War dramatically increased state control of industry, food, and wages. This was reversed after the war ended, but it signalled a 'period of confusion' in the interwar years, in which collectivism vied with capitalism. Under British Prime Minister Stanley Baldwin's leadership from 1924 to 1929, social and economic policies were implemented in order to prevent a class war.

By 1931, with depression, unemployment, strikes, and fear of revolution and the power of the mass, there was a clear concern over liberal laissez-faire governance and an increasing voice calling for state control.[65] Thus this culture of state intervention in more and more aspects of everyday life, along with fears of the mass, can be seen to relate to the shifts in police control of London's streets outlined in this chapter.

Tripp's promotion in 1932 had come just one year after a change at the top of the Metropolitan Police, with the retirement of the 'gentle' commissioner, Julian Byng, and the arrival of the 'autocratic, intolerant' Hugh Trenchard, as David Ascoli has described them.[66] Tripp himself brought a new style. His predecessor, Frank Elliott, had been described as a 'kindly' and 'much-liked' old policeman.[67] Tripp, by contrast, was 'forceful and cogent in conference and a born fighter, he was impatient of compromise and delay and he both gave and took hard knocks'.[68] In many ways, Tripp's 1934 visit to America was symbolic of a harder-edged approach to match the changing times. We are looking at a shift in the characterization of congestion in the early 1930s paralleling a shift in wider relationships between police and public, both in American cities and in London. Tripp's rhetoric was about public safety, not individual liberty as had been paramount in a previous generation. His was more an American (interventive, controlling) police model, replacing an older British one of consent and freedom. And in his report of the 1934 trip to America, we can discern disturbing influences that were to shape his approach.

Alongside Tripp's statistical reporting of accident rates and so on, he offered verbatim accounts of confrontations between local police and the citizenry that throw a frank light on the problems of control in 1930s America. In fact, the report's potentially incendiary nature was recognized by Hugh Trenchard, who, asking for it to be made confidential, observed 'I do not want to see extracts from the report published in every newspaper. I think it would have a very bad effect if some of the sentences in it were taken away from their context.'[69]

Tripp's visit to America came just after the end of the prohibition era. A country-wide ban on alcohol had been in operation from 1920 to 1933 and had led to a breakdown of respect for law-enforcement bodies, among other effects. Cities such as Chicago had become the focus of organized bootlegging gangs operating violent black markets, with Al Capone becoming an iconic figure in American criminal lore, and it is difficult to overstate the negative effects (though in many cases unintended) on American urban culture of this movement.[70]

Tripp was in Chicago just months after prohibition was repealed, and made the following observations about the breakdown of civil society in the city and others like it:

> it must be remembered that in these American cities (Chicago especially) there is a mixed population, much of which is drawn from the dregs of Europe. Of the murders in Chicago about 30% are negroes killing negroes, and, as the Police put it, 'when a negro gets a drop of liquor into him he

kills. He will kill for anything. He will kill for an electric light globe, a cigar, or really for nothing at all.' Another 10% are Italians killing Italians, and the majority are gangsters killing gangsters, or criminals, criminals. The risk to life run by the ordinary white man who behaves himself and manages to keep clean of rackets is much lower than might be inferred from the statistics.[71]

Apart from forcefully articulating the era's typically racialized discourse, what this translated to was a 'bold and rather extreme policy', in Tripp's words, of active policing.[72] He described night-time police cruises in which people deemed suspicious were stopped; searched; and, if still suspicious, arrested and taken to the police station for questioning. His trip was ostensibly to learn about traffic control but, as he later reported, it was these night-time experiences that gave him a thrill. 'I confess that I found it most entertaining', he recalled. 'There was a spice of excitement and adventure about it which was not to be denied.'[73] In one account, he described an episode in the early hours of the morning when the cruiser in which he was riding took a call to attend a robbery-with-violence at a drug store in a working-class African-American neighbourhood. Once on the scene, the police sergeant and Tripp went inside to take down details from the owners, a black couple. Tripp reported candidly what happened:

> We found a negro and negress behind the counter looking very anxious and worried, for they were afraid of being implicated. Particulars were taken. After he had finished, the Sergeant turned to me. 'Gee', he said in a tone of supreme contempt, 'they've only rolled a nigger for a couple of dahlers, and they call it <u>rahbry</u>'.[74]

There was also endemic corruption of both the police and the court system as a result of prohibition. Police corruption took the form of large-scale political interference in the service, individually corrupt officers, corrupt practice in compacts with criminals such as Al Capone, and the overlooking of offences in order to obtain political advantage. The last type was particularly relevant to the traffic service, where widespread traffic offence ticket-fixing took place.

The tension here was about individual liberty versus public order, and it is clear that Tripp had some sympathy with the American approach. He began by remarking as follows:

> one of the main reasons for the results produced in America is that the Police (armed with appropriate laws) contrive to make the motoring public 'toe the line' (so to speak) in a way that we never have succeeded in doing. Speed is <u>lower</u> and <u>more uniform</u> and vehicles keep line better. That effect is not produced by exhortation of Press campaign; it is produced partly by mechanical control, but especially by intensive law enforcement.[75]

This was not about lower individual speeds leading to better journey times across the population of vehicles; it was more fundamental (and less nuanced) than that:

cops want order. Tripp expanded on his theme, stating that the police in all countries had been 'put to the test by the sudden menace' of motor traffic in the first 30 years of the twentieth century, and some had reacted better than others:

> American institutions and methods seem to have reacted more freely and better than our own, which have proved too inflexible and conservative. We appear to have shown less adaptability of mind; and as a result we have by contrast failed to a greater degree. To say this is not of course to decry the British character or to exalt the American by contrast. Undue quickness of reaction may easily lead to instability; the fiasco of Prohibition is a case in point.[76]

Yet the idea of American people submitting to curbs on personal liberty for the greater public good seems as surprising as expecting the British people to do so. Clearly, prohibition is an important concept in understanding Tripp's policing culture. But there was more to it than that. By looking closer at the particular conditions in America in the 1920s and 1930s we can find a more complex story of why, it seems, Chicagoans, New Yorkers, Londoners, and Detroiters reacted to police traffic controls in the ways they did. In particular, and in light of the racist culture of policing witnessed and described by Tripp, we will look at the role of segregation in US cultures of governance.

Segregation and cosmopolitanism

Following his return from America in 1934, Tripp made the following observation about exclusion in the context of traffic in US towns:

> Regulation of traffic is all in favour of the private car; commercial vehicles are rigorously excluded from many of the main traffic arteries in towns, while in the country the parkways, which are specially constructed motor roads, exhibit notices 'No buses, no trucks, no commercial traffic'.[77]

Tripp became a prolific writer on matters relating to traffic and the public. Throughout his work, there is a strong rhetoric of exclusion – as part of a culture of categorization – and it was in the categorization of pedestrians that the language of separation was most clear.

The term Tripp consistently used for the separation of pedestrians from vehicles was 'segregation', giving a new meaning to a term that had been used by his predecessors, from the early 1920s, to mean separating horse-drawn traffic from motor vehicles, or freight and public-service vehicles from private cars.[78] In Tripp's influential 1938 book *Road Traffic and Its Control*, he described two forms of pedestrian separation, namely place-segregation and time-segregation, as follows:

In the case of new roads required for the needs of wheeled traffic, pedestrians can be excluded altogether. In the case of existing roads, footpaths provide the means of segregation, so far as pedestrians proceeding longitudinally are concerned – more especially if guard-rails are added. For pedestrians crossing the road, however, a system of place-segregation (by bridge or subway) is very costly, and the less straightforward plan of time-segregation must generally be invoked.

In the latter case, pedestrians and motor vehicles needed to share the same physical road space but not at the same time, and crossings protected by traffic signals provided the segregating technology. However, care needed to be taken, cautioned Tripp, to balance the needs of pedestrians to cross the road with the need for 'vehicular fluidity' in order to prevent the latter's 'virtual immobility'. And he concluded with the warning that pedestrians would never be fully safe as long as they were 'free to neglect such means of safety as are provided for their benefit'.[79]

He had been presenting the notion of pedestrian segregation since his appointment as Traffic Commissioner. In 1933, in front of the Institute of Transport, he made a bold set of claims about the unfairness of the road scene towards motorists:

> Pedestrians demand the right to use any part of the footway or carriageway at their discretion. The footway is forbidden ground for vehicles; but the carriageway is not forbidden ground for pedestrians ... Vehicles are herded into droves on each side of the road, and are ordered about by police officers and traffic signals and made to use one way streets and so forth; pedestrians are completely free agents.[80]

The solution was that 'steps should be taken wherever possible to segregate vehicular and pedestrian movement, and that in the case of arterial routes, the segregation should be absolute'.[81] This was a solution based on classification, and the apportionment of value to citizens and their use of the road. Tripp concluded:

> The first step towards solution of the whole problem of the design of streets and traffic requirements must be a critical analysis of the demands of the various classes of road users, and a strict assessment of them in order to determine which are legitimate and which excessive and therefore inadmissible.[82]

To what extent might this culture of segregation and exclusion have been influenced by American ideas of race? A clue lies in the concept of 'cosmopolitanism'. While in Chicago, Tripp had asked why there was no jaywalking law prohibiting pedestrians from crossing except at fixed places. He recalled the reply:

The Chicago Police told me that they thought it would be difficult to enforce such a law 'in a cosmopolitan city like Chicago', but they added that a law had been successfully enforced in Minneapolis where the population is more homogeneous.[83]

The term 'cosmopolitan' is telling in this context. At its simplest, the comment may simply have meant that cities with many visitors or 'outsiders' had trouble enforcing rules those visitors were not familiar with. There is also, as Peter Norton has eloquently described, a political history of the concept of jaywalking that may have played a significant role in the judicial reluctance to convict.[84] But there may be a deeper reading of the specific term 'cosmopolitan'. Characterizations of major US cities had for many years been provided by tourist literature and guide books written by English and French authors that, as David Gilbert and Claire Hancock have shown, significantly shaped the European experience of America.[85] In this literature, cosmopolitanism was a distinctly pejorative term, meaning 'monstrous' ethnic diversity, a degenerate state close to the abusive American concept of 'mongrelization', although its meanings were complex and variable.

Tripp was not alone in taking part in a sightseeing tour of Chicago's poor neighbourhoods. 'Slumming' – treating mass immigration as a tourist spectacle – was a standard experience in the average European visitor's itinerary in major US cities, with ethnic neighbourhoods in New York in particular acting as 'a kind of museum of humanity, where the customs of the world could be examined', as Gilbert and Hancock put it, and at its heart, they contend, the conception of the 'cosmopolis' as a 'city polluted by diversity' was racist.[86] Tripp's accounts of New York and Chicago fit well into this culture of touristic slumming and the sordid thrill of what one 1924 guide described as 'motley cosmopolitanism'.[87]

The first four decades of the twentieth century were a period in which racial segregation became normal in the USA, not just in the southern states, which practised extensive and overt spatial segregation in its towns and cities, but also in the northern ones, which encouraged the deliberate ghettoization of urban neighbourhoods and the legislation of 'whites only' restrictions in proliferating New Deal suburbs.[88] In reporting his experiences in Chicago and other American cities, as we have seen, Tripp described local police forces that were consciously aware of – and prejudiced negatively against – minority groups. Black people were 'negroes' and 'negresses' or, just as commonly, referred to by another well-known but persistently racializing and humiliating epithet. Black men, police infamously held, were likely to be rapists, and working-class white women were fair game; both were assumed to be sexually promiscuous.[89] Homosexual men were designated 'fairies' or 'pansies', both popular contemporaneous derogations.[90]

For the Chicago police, the local population was an assemblage of distinct identifiable groups – *ethnicities* and *races* – to be categorized and controlled.[91] But segregationist ideas were not restricted to America: they were growing in the

UK at this time too, and, by reflecting on the mutability of the concept, we can start to understand the context in which Tripp chose to shape London's traffic scene by segregation. There is no suggestion that *racial* segregation was implied or considered by Tripp in his proposals for London's streets. East India Dock Road was not singled out for guard rails owing to a particular ethnicity of the local population. It was simply the location for the first installation of what Tripp hoped would be a universal technology across the capital, from Whitechapel to Knightsbridge, and from Commercial Road to Oxford Street. But it is nevertheless notable that he chose to use the term 'segregation' consistently, and he was certainly operating in a segregationist culture of public governance that viewed the classification of people along ethnic, gender, and sexual lines as normal.

Paul Rich has examined the situation in Britain in the early twentieth century, decades before the emerging racial conflict following the Second World War and the 'new racism' of the 1970s. Describing segregation in the USA at that time, he observes that it:

> fell substantially into the arena of mainstream political discourse of 'middle opinion', and ... for a significant period from the 1890s up to the 1940s the concept seemed eminently reasonable even to people of moderate and liberal political persuasion.

Rich notes the 'international nature' of segregation and the 'readily exportable' nature of the term, going on to explore its meanings in Britain.[92] Susan Smith has also written on the politics of race in the UK in this period, describing how ideologies of racism, nationalism, and segregationism express a complex network of beliefs.[93]

We can observe a nationalism in Tripp's rhetoric of the London police with respect to non-British forces, and this becomes racially inflected, not just when words such as 'cosmopolitan' are deployed, but also in his extensive negative descriptions of racial and cultural diversity. At the local London level, segregationist ideas – the geographical allocation of space according to culturally constructed categories such as 'pedestrian' and 'motorist' – can be seen to be heavily imbued with gender, class, race, age, and other embedded prejudices. Thus, although not necessarily overtly racist, we can see in Tripp's localized pedestrian segregation/exclusion an expression of a global movement – what Phillip Gordon Mackintosh has termed 'infrastructural racism', where the 'uncritical organization of public infrastructure abet[ted] and affirm[ed] Victorian and Edwardian conceptions of race'.[94] Importantly, for Smith, such thinking may not have been 'a rigid formula transported between continents, but a system of meanings, norms and expressions, moulding and adjusting to new challenges and changed circumstances'.[95]

In this context, it is hard to read a comment made by Tripp in 1936, expanding on his theme of pedestrian segregation in London, without hearing echoes of the American racial debate. This time he was introducing stronger notions of dominance, coercion, and promiscuity, as follows:

The principle to be aimed at must be to give the wheeled traffic a defined dominance over pedestrian traffic in all roads of an arterial character, but at the same time to provide regulated zones of safety for the pedestrians. In addition, in order to prevent promiscuous crossing by pedestrians, it may be necessary, as I have long advocated, to provide guard rails along main roads ... If such facilities and safeguards are provided it will pay both vehicular traffic and pedestrians to conform to orderly movement, and they will do so without any form of coercion by the police.[96]

One might substitute 'whites' for 'wheeled traffic', and Tripp's term, 'negroes', for 'pedestrians' and hear a culture of racial categorization and exclusion common to both contexts. Yet there may be a further link between traffic infrastructure and segregation in the notion of cordons, both physical and social. By examining the origins of segregationist concepts in the European colonialist context we can discern ideas linking race to health, injury, death, and disease through material technologies.

Promiscuity, disease, impurity, and 'cordoning off'

Carl Nightingale, following the work of Daniel Headrick, has plotted a racialization and shift in the meaning of segregation in the 1890s following outbreaks of plague in Hong Kong and India.[97] In response, city public health officials began removing infected citizens to hospitals or ships. 'This technique', Nightingale describes, 'they called isolation, or, increasingly, segregation. Race mattered: the main goal was to protect Europeans, and local Asian people got the roughest treatment.'[98] The plague and its segregationist control response quickly spread to America, Europe, and Africa, adding to local cultures of disease, vector, and fear. In Dakar, Headrick observes, colonial officials enforced a rigid segregation of Europeans from Africans: 'Quickly the idea spread in the administration and in the white community that *all* Africans were suspect and should be evacuated.'[99]

Moreover, there was an existing colonial background of the building of expensive infrastructure to reinforce segregation and public health control – sewers, water supply, and building regulations.[100] Stephen Legg has discussed the case of hygiene and urban housing congestion in colonial Delhi in the 1930s, and governmental responses to the problem in relation to local populations.[101] Class blended with race in these infrastructural interventions. Describing colonial cities, Headrick explains as follows:

Here, among Europeans and those few who could afford to emulate them, a growing concern for hygiene provided the motive for segregation, while new technologies of water supply, sewage disposal, and building materials provided the means. And since these technologies were costly, they were reserved for the better-off.[102]

Legg has also explored the use of spatial segregation to control prostitution in India in the first decades of the twentieth century, demonstrating that local cultures of governance intertwined notions of race with gender, class, homosexuality, cosmopolitanism, and promiscuity through disease – in this case venereal – with segregation a crucial method of control. His argument, that the Indian colonial state had embedded within it 'a spatial and governmental mindset that assumed segregation was the most effective means of controlling both the prostitute and ... the military', helps shed light on the arguably related situation of 'promiscuous pedestrians' mixing with modern vehicles on London's streets.[103]

Anne McClintock has gone further. Examining the 'degeneration' trope of the late nineteenth century, she observes the following:

> In the metropolis, the idea of racial deviance was evoked to police the 'degenerate' classes – the militant working class, the Irish, Jews, feminists, gays and lesbians, prostitutes, criminals, alcoholics and the insane – who were collectively figured as racial deviants, atavistic throwbacks to a primitive moment in human prehistory, surviving ominously in the heart of the modern, imperial metropolis.[104]

She goes on to explain the social purpose of such characterization, stating that 'social classes or groups were described with telling frequency as "races", "foreign groups", or "nonindigenous bodies", and could thus be cordoned off as biological and "contagious", rather than as social groups'.[105] This cordoning-off activity she considers part of Julia Kristeva's notion of 'abjection', the process by which a social being creates itself by expelling 'impure' elements, which never truly disappear but 'haunt the edges of the subject's identity with the threat of disruption or even dissolution'. Crucially, she explains that 'Certain threshold zones become abject zones and are policed with vigour: the Arab Casbah, the Jewish ghetto, the Irish slum, the Victorian garret and Kitchen, the squatter camp, the mental asylum, the red light district, and the bedroom.'[106]

In 1930s London, we can consider a new 'abject people' and their 'abject zone': pedestrians on the pavements who, in Tripp's characterization, were racialized, gendered, sexualized, aged, and classed. They formed a group whose disordered presence in the roadway was slowing those people who were driving through the streets. That was their contagion, in the sense of a pollutant or a corrupting influence: pedestrians obstructed the passage of vehicle drivers in a hurry, breaking down Tripp's perceived proper order of the road. Yet pedestrians could only be marginalized, not eradicated, as even the driver was a pedestrian before getting into the car.

Ideas about segregational infrastructure were thus in the air in the 1920s and 1930s, and the connection between traffic segregation and injury, death, and disease was expressed repeatedly in Tripp's writings and the pronouncements of his peers. At one 1932 public conference attended by Tripp, Sir Henry Pigott, Deputy Secretary to the Ministry of Transport, said that:

every day, on an average, 18 people were killed and some 550 injured, many of them maimed for life, on the roads of Great Britain. If some disease or plague were to claim victims on that scale the whole medical faculty would be mobilized to meet the situation and provide a remedy.[107]

Tripp opened his own 1938 book with the notion of motor traffic 'beginning to strangle its own circulation', asking for pedestrian segregation from roadways to rectify the 'inevitable and automatic' effect of non-segregation, namely 'death and injury', although he appreciated that the situation was probably inevitable, as to have placed restrictions on motor traffic sooner 'would have thrown upon the infant enterprise a heavier burden than it could bear, and would perhaps have stifled it at birth'.[108] The influential 1939 report of the House of Lords select committee on road accidents, to which Tripp substantially contributed and which Joe Moran has observed was regarded as a motorist's charter, was itself structured in two parts – 'The Disease' and 'The Remedy' – where the first section in the latter part was entitled 'Segregation Must Come'.[109] Tripp himself had provided the committee with a detailed report that included a description of the 'cure' needed for the traffic problem.[110]

It therefore seemed natural that physical, spatial segregation was an appropriate response to death and disease, and whilst the intensity of disease-spread segregation had receded by 1920, it was replaced by the emerging culture of urban planning. As Nightingale has described:

> For the new breed of comprehensive city planner who led this movement, the connection between sanitation and segregation was the stuff not of fear or disgust, but of the grandest hopes of Western empire and civilization. From this exalted inspiration arose the colonial era's most extravagant monuments to urban segregation.[111]

Amid all this international rhetoric of public health and the enforcement of spatial segregation through infrastructure, we can see that the response to the disease crisis mirrored closely Alker Tripp's response to the epidemic of injury and death on London's shared road space as motorization took hold: different groups should be isolated from one another through the built form of the city. This was expressed in two ways. The first was in Tripp's plans to reshape the relationship between pedestrians and motorists on London's streets using the control technologies of guard rails and traffic lights. But Tripp's more widespread expression of spatial segregation came from his influence over generations of urban planners – which we still live with today.

Conclusions

The Metropolitan Police's approach to traffic congestion and circulation in the nineteenth century can be characterized as liberal, reactive, reluctant, consensual, localized, and fragmented. Yet the first three decades of the twentieth century saw this approach shift, becoming increasingly interventionist, proactive, holistic, coordinated, and planned. This mirrored British society's journey from laissez-faire to liberal democratic. In many ways, the Metropolitan Police in the 1930s, under Traffic Commissioner Alker Tripp, was keener for change on the streets than one might have thought, given its conservative history. On the surface, we can see this partly as a power struggle between the police and other professional groups over who controls the streets and the movement that takes place in them. Two critical relationships in the case of the Metropolitan Police were with the engineers and scientists at the Ministry of Transport, and with the emerging town planning profession. But we can also see this as the circulation of ideas on policing and social control among major cities around the world. Thus concepts originating in American cultures of prohibition and racial tension, or Indian notions of sexual promiscuity, impurity, and disease, found new meanings in London's particular circumstances.

Partly these circumstances relate to London's structures, including its ancient non-gridded layout and its complex layers of local governance. These structures live long and assert dominance. In the case of layout, Tripp sought to introduce grid-like traffic behaviour: a gridded structure offers more opportunities to reduce congestion, as traffic can take many different routes between two points – there is much redundancy – whereas London's traffic can easily be halted by blockages at a handful of key junctions or river crossings. Sir Christopher Wren failed in the seventeenth century to change the layout, with grid historian Hannah Higgins offering clues as to why, remarking that 'the gridiron is an expression of a highly regulated, tightly administered culture'.[112] This demonstrates the connection to Tripp's second problem, that of a lack of centralized governance and a single traffic body for London. Both its physical layout and its political governance express a liberal society and, despite the sweeping changes of the twentieth century, cities change more slowly than one might think.

But the reception and transformation of ideas such as segregation from one place and culture to another are not just shaped by structures such as layouts and governance. Individuals in the network matter too. What is clear is that Tripp was not just a traffic policeman. He was also an engineer and planner as well as an enthusiastic campaigner and politician. As a figure moving between the margins and the mainstream of planning, his ideas came to the attention of planners who were sensitive to visions of holistic control, and they took root. Long after Tripp's retirement in 1947, his notions of segregation had helped shape a generation of planners to view pedestrians as an impurity to be isolated – to be marginalized. But this account is a messier reality than conventional ideas of powerful top-down state control and planning might suggest – such ideas were challenged by London's physical and historical reality as well as by the actions of individuals from councillors to shopkeepers seeking to preserve their place in the network. Embedded infrastructures (in the socio-politico-technical sense) have long lives, and control of London's traffic has always been about retrofitting the nineteenth-century city, with only partial success.

Notes

1 A modified version of this chapter was published as David Rooney, 'Keeping Pedestrians in Their Place: Technologies of Segregation on the Streets of East London', in *Architectures of Hurry: Mobilities, Cities and Modernity*, ed. Phillip Gordon Mackintosh, Richard Dennis, and Deryck Holdsworth (Abingdon: Routledge, 2018), 120–36.

2 *The Times*, 29 May 1936, 11.

3 The National Archives (hereafter TNA), MEPO 2/6748, memorandum, Divisional Road Engineer (London) to Chief Engineer, 17 August, 1935.

4 *The Times*, 2 September 1936, 8.

5 Tower Hamlets Local History Library and Archives, 'Photograph Library: Street Scenes: East India Dock Road', P02766.

6 See for instance his paper to the Institution of Electrical Engineers on 13 November 1933, reported in *The Times*, 14 November 1933, 11.

7 TNA, MEPO 2/6748, file note by Tripp, probably sent to Arthur Dixon (Home Office), 3 May, 1935.

8 Ministry of Transport, *Second Annual Report of the London and Home Counties Traffic Advisory Committee* (London: HMSO, 1926), 38; *The Times*, 7 April 1934, 10; 11 March 1935, 9; Alker Tripp, 'Progress in Road Safety Measures', *Police Journal* 11 (1938): 442; Muhammad Ishaque and Robert Noland, 'Making Roads Safe for Pedestrians or Keeping Them out of the Way? An Historical Perspective on Pedestrian Policies in Britain', *Journal of Transport History* 27, no. 1 (2007): 115–37; Richard Hornsey, '"He who Thinks, in Modern Traffic, Is Lost": Automation and the Pedestrian Rhythms of Interwar London', in *Geographies of Rhythm: Nature, Place, Mobilities and Bodies*, ed. Tim Edensor (Farnham: Ashgate, 2010), 99–112; and Keith Laybourn and David Taylor, *Policing in England and Wales, 1918–39: The Fed, Flying Squads and Forensics* (Basingstoke: Palgrave Macmillan, 2011), Chapter 6.

9 Ministry of Transport, *Seventh Annual Report of the London and Home Counties Traffic Advisory Committee* (London: HMSO, 1931), 27; Ministry of Transport, *Sixth Annual Report of the London and Home Counties Traffic Advisory Committee* (London: HMSO, 1930), 22; Ministry of Transport, *Second Annual Report*, 30–1; J. D. Francis, 'The First Traffic Lights in Britain', *The Signalling Record: The Journal of the Signalling Record Society* 91 (February 2002): 5–15. There was also an experiment in 1868 using a gas-lit installation that was discontinued following an accident; see James Winter, *London's Teeming Streets, 1830–1914* (London: Routledge, 1993), 34–6; and Carlos Lopez Galviz, 'Mobilities at a Standstill: Regulating Circulation in London c. 1863–1870', *Journal of Historical Geography* 42 (2013): 73–6.

10 On traffic lights and signals, see for instance G. A. H. Wootton, 'Mechanical Aids for Traffic Control and Other Police Purposes', *Police Journal* 1 (1928): 89–97; H. E. Aldington, 'Traffic Control: Comparative Methods', *Police Journal* 2 (1929): 193–210; Henry Watson, *Street Traffic Flow* (London: Chapman and Hall, 1933), Chapters 9–10; and Alker Tripp, *Road Traffic and Its Control* (London: E. Arnold, 1938), Chapter 12.

11 British Pathé, *Trafalgar Square: 70,000 Vehicles Daily and Not One Traffic Policeman!* (film), 1933.

12 TNA, MEPO 2/6748, letter, Tripp to Arthur Dixon (Home Office), 3 May 1935.

13 TNA, MEPO 2/6748, letter, Tripp to Archibald Matheson (Ministry of Transport), 30 April 1935.

14 TNA, MEPO 2/6748, letter, Tripp to Archibald Matheson (Ministry of Transport), 30 April 1935.

15 On cultures of techno-social system building, see Thomas P. Hughes, *Networks of Power: Electrification in Western Society, 1880–1930* (Baltimore: Johns Hopkins University Press, 1983).

16 TNA, MEPO 2/6748, note, Tripp to Select Committee, n.d. but about November 1938.

17 TNA, MEPO 2/6748, letter, Tripp to Hugh Trenchard, 9 March 1935.

18 TNA, MEPO 2/6748, 'Memorandum on Automatic Traffic Signals – Comprehensive Scheme for London', for Select Committee on the Prevention of Road Accidents, by Alker Tripp, 30 November 1938; *The Times*, 24 October 1935, 19; Tower Hamlets Local History Library and Archives, L/PMB/A/1/36, meeting, 23 January 1936, in 'Poplar Borough Council, Minutes, 1935–6'; L/PMB/A/10/17, meeting, 17 February 1936, in 'Poplar Borough Council, Works Committee, Minutes, February 1936– January 1937'.

19 *Manchester Guardian*, 11 March 1935, 8.

20 *Manchester Guardian*, 29 May 1936, 10.

21 On the position of pro- and anti-motoring lobbies regarding pedestrians in this period, see Michael John Law, *The Experience of Suburban Modernity: How Private Transport Changed Interwar London* (Manchester: Manchester University Press, 2014), Chapter 9; and on age, class, and gender issues surrounding pedestrian crossings, see Joe Moran, 'Crossing the Road in Britain 1931–1976', *Historical Journal* 49, no. 2 (2006): 477–96.

22 *The Times*, 15 January 1935, 14.

23 *Pedestrians' Association Occasional News Letter*, October 1934, 5–6.

24 *Pedestrians' Association Quarterly News Letter*, October 1935, 9.

25 Living Streets Archive, Minutes of the Pedestrians' Association, 14 January 1936; *Pedestrians' Association Quarterly News Letter*, January 1936, 6–7; *The Times*, 26 November 1935, 11; and on the interest-group politics of road safety in the 1930s, see William Plowden, *The Motor Car and Politics in Britain 1896–1970* (London: Bodley Head, 1971), Chapter 13.

26 Pedestrians' Association, *Ninth Annual Report of the Pedestrians' Association* (London: Pedestrians' Association, 1937), 6–7.

27 Pedestrians' Association, *Tenth Annual Report of the Pedestrians' Association* (London: Pedestrians' Association, 1938), 5.

28 Anon., *The Post Office London Directory for 1936* (London: Kelly's Directories, 1936), 388–9; see also Stephen Porter, ed., *Survey of London*, Vol. XLIII: *Poplar, Blackwall and the Isle of Dogs* (London: Athlone Press for the Royal Commission on the Historical Monuments of England, 1994), 120–70.

29 *East London Advertiser*, 30 May 1936, 2.

30 TNA, MEPO 2/6748, memorandum by Tripp, probably 1936.

31 TNA, MEPO 2/6748, minute, Tripp to Game, 10 January 1938.

32 Hornsey, 100.

33 On the history of traffic policing, see Laybourn and Taylor, Chapters 5–8.

34 Charles Stewart Murdoch, 'Appendix F: The Regulation of Traffic in the Metropolis and City of London by the Police', in Royal Commission on London Traffic, *Appendices to the Report of the Royal Commission on London Traffic (Volume IV)* (London: HMSO, 1906), 642.

35 Winter, Chapters 3–4.

36 Winter, 52.

37 Winter, 60–1.

38 Winter, 41.

39 Winter, 42–9.

40 Galviz, 66.

41 Galviz, 62.

42 House of Lords, *Report by the Select Committee of the House of Lords on the Prevention of Road Accidents [Session 1938–39]: Together with the Proceedings of the Committee, Minutes of Evidence and Index* (London: HMSO, 1939), 54.

43 K. Rivers, *History of the Traffic Department of the Metropolitan Police* (London: Metropolitan Police, 1969); Martin Fido and Keith Skinner, *The Official Encyclopedia of Scotland Yard* (London: Virgin, 1999), 265–6.

44 See *The Times*, 18 January 1926, 7; and 19 January 1926, 7.

45 Clive Emsley, '"Mother, What Did Policemen Do When There Weren't Any Motors?" The Law, the Police and the Regulation of Motor Traffic in England, 1900–1939', *Historical Journal* 36, no. 2 (June 1993): 357.

46 Plowden, *The Motor Car and Politics in Britain*, Chapters 6, 11, 12, 13.

47 Laybourn and Taylor, 9, 106.

48 Alker Tripp, 'Police and Public: A New Test of Police Quality', *Police Journal* 1 (1928): 533.

49 TNA, MEPO 2/6748, file note by Tripp, probably sent to Arthur Dixon (Home Office), 3 May 1935.

50 Miles Ogborn, 'Traffic-Lights', in Steve Pile and Nigel Thrift, *City A–Z* (London: Routledge, 2000), 263–4.

51 Christopher Otter, 'Cleansing and Clarifying: Technology and Perception in Nineteenth-Century London', *Journal of British Studies* 43, no. 1 (January 2004): 63; see also Christopher Otter, 'Making Liberalism Durable: Vision and Civility in the Late Victorian City', *Social History* 27, no. 1 (January 2002): 1–15; and Chris Otter, 'Making Liberal Objects: British Techno-Social Relations 1800–1900', *Cultural Studies* 21, nos. 4–5 (September 2007): 570–90.

52 Patrick Joyce, *The Rule of Freedom: Liberalism and the Modern City* (London: Verso, 2003), Chapters 2, 5, 6 (quotation on 262).

53 Galviz, 64.

54 Thomas Osborne and Nikolas Rose, 'Governing Cities: Notes on the Spatialisation of Virtue', *Environment and Planning D: Society and Space* 17 (1999): 744.

55 On the cultural meaning of the American city grid system to visitors from London in this period, see David Gilbert and Claire Hancock, 'New York City and the Transatlantic Imagination: French and English Tourism and the Spectacle of the Modern Metropolis, 1893–1939', *Journal of Urban History* 33, no. 1 (November 2006): 92–3.

56 TNA, MEPO 2/6748, memorandum, Divisional Road Engineer (London) to Chief Engineer, 17 August 1935.

57 TNA, MEPO 2/5937, *Report by H. Alker Tripp, Assistant Commissioner, on Visit to America, October*, 1934, III:2.

58 TNA, MEPO 2/5937, III:11–12.

59 TNA, MEPO 2/5937, III:2–3.

60 TNA, MEPO 2/5937, III:3.

61 TNA, MEPO 2/5937, III:4.

62 TNA, MEPO 2/5937, III:5.

63 TNA, MEPO 2/5937, III:6.

64 Martin Pugh, *State and Society: A Social and Political History of Britain 1870–1997*, 2nd edn (London: Arnold, 1999), 62.

65 Pugh, Parts II and III, and in particular 120–1 and 216–8; see also the essays in Mary Langan and Bill Schwarz, eds., *Crises in the British State 1880–1930* (London: Hutchinson, 1985).

66 David Ascoli, *The Queen's Peace: The Origins and Development of the Metropolitan Police 1829–1979* (London: Hamish Hamilton, 1979), 219, 227.

67 *The Times*, 27 March 1939, 14.

68 *The Times*, 23 December 1954, 9.

69 TNA, MEPO 2/5937, memorandum, Trenchard to Tripp, 18 December 1934.

70 On the history of the American Prohibition, including its effects, see for instance Norman Clark, *Deliver Us from Evil: An Interpretation of American Prohibition* (New York: Norton, 1976).

71 TNA, MEPO 2/5937, *Report by H. Alker Tripp on Visit to America*, II:6.
72 TNA, MEPO 2/5937, II:7.
73 TNA, MEPO 2/5937, IV:1.
74 TNA, MEPO 2/5937, IV:7.
75 TNA, MEPO 2/5937, IIA:5.
76 TNA, MEPO 2/5937, III:1.
77 TNA, MEPO 2/5937, I:7.
78 See for instance Chief Constable Arthur Bassom in *The Times*, 4 November 1924, 11.
79 Tripp, *Road Traffic and Its Control*, 136–7.
80 Alker Tripp, 'The Design of Streets for Traffic Requirements', *Journal of the Institute of Transport* 15, no. 2 (December 1933): 76.
81 Tripp, 'The Design of Streets', 80.
82 Tripp, 'The Design of Streets', 84.
83 TNA, MEPO 2/5937, *Report by H. Alker Tripp on Visit to America*, I:8.
84 See Peter Norton, 'Street Rivals: Jaywalking and the Invention of the Motor Age Street', *Technology and Culture* 48 (2007): 331–59; and Peter Norton, *Fighting Traffic: The Dawn of the Motor Age in the American City* (Cambridge, MA: MIT Press, 2008).
85 Gilbert and Hancock.
86 Gilbert and Hancock, 96–8.
87 'Rider's New York City' (1924), quoted in Gilbert and Hancock, 87.
88 See Carl Nightingale, *Segregation: A Global History of Divided Cities* (Chicago: University of Chicago Press, 2012), Chapter 10; David Freund, *Colored Property: State Policy and White Racial Politics in Suburban America* (Chicago: University of Chicago Press, 2007); Craig Steven Wilder, *A Covenant with Color: Race and Social Power in Brooklyn* (New York: Columbia University Press, 2000).
89 TNA, MEPO 2/5937, *Report by H. Alker Tripp on Visit to America*. See for instance IV:8, where Tripp described a white woman passed out in the street from alcohol: 'I enquired, as we drove off, whether being as tight as that in the Black quarter some negro wouldn't have her before the morning. "Yeah", was the unconcerned reply, "and some of them white women come on purpose for it".'
90 TNA, MEPO 2/5937, IV:9; See George Chauncey, *Gay New York: Gender, Urban Culture, and the Making of the Gay Male World, 1890–1940* (New York: Basic, 1994).
91 On race and social control see, for example, Matthew Frye Jacobson, *Whiteness of a Different Color: European Immigrants and the Alchemy of Race* (Cambridge, MA: Harvard University Press, 1998); Wilder; David Roediger, *Working toward Whiteness: How America's Immigrants Became White. The Strange Journey from Ellis Island to the Suburbs* (New York: Basic, 2005).
92 Paul Rich, 'Doctrines of Racial Segregation in Britain: 1900–1944', *New Community: Journal of the Commission for Racial Equality* 12, no. 1 (Winter 1984): 77.
93 Susan Smith, *The Politics of 'Race' and Residence: Citizenship, Segregation and White Supremacy in Britain* (Cambridge: Polity Press, 1989), Chapter 1 (esp. 4–7).
94 Phillip Gordon Mackintosh, 'The "Occult Relation between Man and the Vegetable": Transcendentalism, Immigrants, and Park Planning in Toronto, *c.* 1900', in *Rethinking the Great White North: Race, Nature and the Historical Geographies of Whiteness in Canada*, ed. Andrew Baldwin, Laura Cameron, and Audrey Kobayashi (Vancouver: UBC Press, 2011), 105.
95 Smith, 13.
96 Alker Tripp, 'The Traffic Problem', *Police Journal* 9 (1936): 90.
97 Nightingale, Chapter 6; Daniel Headrick, *The Tentacles of Progress: Technology Transfer in the Age of Imperialism, 1850–1940* (Oxford: Oxford University Press, 1988), Chapter 5.

98   Nightingale, 159.
99   Headrick, 163.
100 Nightingale, 165.
101 Stephen Legg, 'Governmentality, Congestion and Calculation in Colonial Delhi', *Social and Cultural Geography* 7, no. 5 (2006): 709–29.
102 Headrick, 147.
103 Stephen Legg, 'Stimulation, Segregation and Scandal: Geographies of Prostitution Regulation in British India, between Registration (1888) and Suppression (1923)', *Modern Asian Studies* 46, no. 6 (November 2012): 1462.
104 Anne McClintock, *Imperial Leather: Race, Gender and Sexuality in the Colonial Contest* (New York: Routledge, 1995), 43.
105 McClintock, 48.
106 McClintock, 71–2.
107 *The Times*, 5 May 1932, 11.
108 Tripp, *Road Traffic and Its Control*, 1, 5.
109 House of Lords; Moran, 483.
110 TNA, MEPO 2/6719, *Select Committee on the Prevention of Road Accidents: Memorandum of the Evidence to Be Given on Behalf of the Commissioner of Police of the Metropolis by H. Alker Tripp, CBE, Assistant Commissioner of Police* (1938), 40.
111 Nightingale, 191.
112 Hannah B. Higgins, *The Grid Book* (Cambridge, MA: MIT Press, 2009), 52; on the Wren plan (and others) following the London fire, see for instance Michael Hebbert, *London: More by Fortune than Design* (Chichester: John Wiley & Sons, 1998), 23–9.

Bibliography

Living Streets Archive: Minutes of the Pedestrians' Association.

The National Archives: Metropolitan Police Papers.

The National Archives: Ministry of Transport Papers.

Tower Hamlets Local History Library and Archives: Photograph Library; Poplar Borough Council Papers.

East London Advertiser.

Manchester Guardian.

Pedestrians' Association Occasional News Letter.

Pedestrians' Association Quarterly News Letter.

The Times.

Aldington, H. E. 'Traffic Control: Comparative Methods'. *Police Journal* 2 (1929): 193–210.

Anon. *The Post Office London Directory for 1936*. London: Kelly's Directories, 1936.

Ascoli, David. *The Queen's Peace: The Origins and Development of the Metropolitan Police 1829–1979*. London: Hamish Hamilton, 1979.

British Pathé. *Trafalgar Square: 70,000 Vehicles Daily and Not One Traffic Policeman!* (film), 1933.

Chauncey, George. *Gay New York: Gender, Urban Culture, and the Making of the Gay Male World, 1890–1940*. New York: Basic, 1994.

Clark, Norman. *Deliver Us from Evil: An Interpretation of American Prohibition*. New York: Norton, 1976.

Emsley, Clive. '"Mother, What Did Policemen Do When There Weren't Any Motors?" The Law, the Police and the Regulation of Motor Traffic in England, 1900–1939'. *Historical Journal* 36, no. 2 (June 1993): 357–81.

Fido, Martin, and Keith Skinner. *The Official Encyclopedia of Scotland Yard.* London: Virgin, 1999.

Francis, J. D. 'The First Traffic Lights in Britain'. *The Signalling Record: The Journal of the Signalling Record Society* 91 (February 2002): 5–15.

Freund, David. *Colored Property: State Policy and White Racial Politics in Suburban America.* Chicago: University of Chicago Press, 2007.

Galviz, Carlos Lopez. 'Mobilities at a Standstill: Regulating Circulation in London *c.* 1863–1870. *Journal of Historical Geography* 42 (2013): 62–76.

Gilbert, David, and Claire Hancock. 'New York City and the Transatlantic Imagination: French and English Tourism and the Spectacle of the Modern Metropolis, 1893–1939'. *Journal of Urban History* 33, no. 1 (November 2006): 77–107.

Headrick, Daniel. *The Tentacles of Progress: Technology Transfer in the Age of Imperialism, 1850–1940.* Oxford: Oxford University Press, 1988.

Hebbert, Michael. *London: More by Fortune than Design.* Chichester: John Wiley & Sons, 1998.

Higgins, Hannah B. *The Grid Book.* Cambridge, MA: MIT Press, 2009.

Hornsey, Richard. '"He who Thinks, in Modern Traffic, Is Lost": Automation and the Pedestrian Rhythms of Interwar London'. In *Geographies of Rhythm: Nature, Place, Mobilities and Bodies,* ed. Tim Edensor, 99–112. Farnham: Ashgate, 2010.

House of Lords. *Report by the Select Committee of the House of Lords on the Prevention of Road Accidents [Session 1938–39]: Together with the Proceedings of the Committee, Minutes of Evidence and Index.* London: HMSO, 1939.

Hughes, Thomas P. *Networks of Power: Electrification in Western Society, 1880–1930.* Baltimore: Johns Hopkins University Press, 1983.

Ishaque, Muhammad, and Robert Noland. 'Making Roads Safe for Pedestrians or Keeping Them out of the Way? An Historical Perspective on Pedestrian Policies in Britain'. *Journal of Transport History* 27, no. 1 (2007): 115–37.

Jacobson, Matthew Frye. *Whiteness of a Different Color: European Immigrants and the Alchemy of Race.* Cambridge, MA: Harvard University Press, 1998.

Joyce, Patrick. *The Rule of Freedom: Liberalism and the Modern City.* London: Verso, 2003.

Langan, Mary, and Bill Schwarz, eds. *Crises in the British State 1880–1930.* London: Hutchinson, 1985.

Law, Michael John. *The Experience of Suburban Modernity: How Private Transport Changed Interwar London.* Manchester: Manchester University Press, 2014.

Laybourn, Keith, and David Taylor. *Policing in England and Wales, 1918–39: The Fed, Flying Squads and Forensics.* Basingstoke: Palgrave Macmillan, 2011.

Legg, Stephen. 'Governmentality, Congestion and Calculation in Colonial Delhi'. *Social and Cultural Geography* 7, no. 5 (2006): 709–29.

———. 'Stimulation, Segregation and Scandal: Geographies of Prostitution Regulation in British India, between Registration (1888) and Suppression (1923)'. *Modern Asian Studies* 46, no. 6 (November 2012): 1459–505.

Mackintosh, Phillip Gordon. 'The "Occult Relation between Man and the Vegetable": Transcendentalism, Immigrants, and Park Planning in Toronto, *c.* 1900'. In *Rethinking the Great White North: Race, Nature and the Historical Geographies of Whiteness in Canada,* ed. Andrew Baldwin, Laura Cameron, and Audrey Kobayashi, 85–106. Vancouver: UBC Press, 2011.

McClintock, Anne. *Imperial Leather: Race, Gender and Sexuality in the Colonial Contest.* New York: Routledge, 1995.

Ministry of Transport. *Second Annual Report of the London and Home Counties Traffic Advisory Committee*. London: HMSO, 1926.

———. *Seventh Annual Report of the London and Home Counties Traffic Advisory Committee*. London: HMSO, 1931.

———. *Sixth Annual Report of the London and Home Counties Traffic Advisory Committee*. London: HMSO, 1930.

Moran, Joe. 'Crossing the Road in Britain 1931–1976'. *Historical Journal* 49, no. 2 (2006): 477–96.

Murdoch, Charles Stewart. 'Appendix F: The Regulation of Traffic in the Metropolis and City of London by the Police'. In Royal Commission on London Traffic, *Appendices to the Report of the Royal Commission on London Traffic (Volume IV)*. London: HMSO, 1906.

Nightingale, Carl. *Segregation: A Global History of Divided Cities*. Chicago: University of Chicago Press, 2012.

Norton, Peter. *Fighting Traffic: The Dawn of the Motor Age in the American City*. Cambridge, MA: MIT Press, 2008.

———. 'Street Rivals: Jaywalking and the Invention of the Motor Age Street'. *Technology and Culture* 48, no. 2 (April 2007): 331–59.

Ogborn, Miles. 'Traffic-Lights'. In Steve Pile and Nigel Thrift, *City A–Z*, 262–4. London: Routledge, 2000.

Osborne, Thomas, and Nikolas Rose. 'Governing Cities: Notes on the Spatialisation of Virtue'. *Environment and Planning D: Society and Space* 17 (1999): 737–60.

Otter, Chris. 'Making Liberal Objects: British Techno-Social Relations 1800–1900'. *Cultural Studies* 21, nos. 4–5 (September 2007): 570–90.

Otter, Christopher. 'Cleansing and Clarifying: Technology and Perception in Nineteenth-Century London'. *Journal of British Studies* 43, no. 1 (January 2004): 40–64, 157–60.

———. 'Making Liberalism Durable: Vision and Civility in the Late Victorian City'. *Social History* 27, no. 1 (January 2002): 1–15.

Pedestrians' Association. *Ninth Annual Report of the Pedestrians' Association*. London: Pedestrians' Association, 1937.

———. *Tenth Annual Report of the Pedestrians' Association*. London: Pedestrians' Association, 1938.

Plowden, William. *The Motor Car and Politics in Britain 1896–1970*. London: Bodley Head, 1971.

Porter, Stephen, ed. *Survey of London*, Vol. XLIII: *Poplar, Blackwall and the Isle of Dogs*. London: Athlone Press for the Royal Commission on the Historical Monuments of England, 1994.

Pugh, Martin. *State and Society: A Social and Political History of Britain 1870–1997*, 2nd edn. London: Arnold, 1999.

Rich, Paul. 'Doctrines of Racial Segregation in Britain: 1900–1944'. *New Community: Journal of the Commission for Racial Equality* 12, no. 1 (Winter 1984): 75–88.

Rivers, K. *History of the Traffic Department of the Metropolitan Police*. London: Metropolitan Police, 1969.

Roediger, David. *Working toward Whiteness: How America's Immigrants Became White. The Strange Journey from Ellis Island to the Suburbs*. New York: Basic, 2005.

Rooney, David. 'Keeping Pedestrians in Their Place: Technologies of Segregation on the Streets of East London'. In *Architectures of Hurry: Mobilities, Cities and Modernity*, ed. Phillip Gordon Mackintosh, Richard Dennis, and Deryck Holdsworth, 120–36. Abingdon: Routledge, 2018.

Smith, Susan. *The Politics of 'Race' and Residence: Citizenship, Segregation and White Supremacy in Britain*. Cambridge: Polity Press, 1989.

Tripp, Alker. 'The Design of Streets for Traffic Requirements'. *Journal of the Institute of Transport* 15, no. 2 (December 1933): 73–97.

———. 'Police and Public: A New Test of Police Quality'. *Police Journal* 1 (1928): 529–39.

———. 'Progress in Road Safety Measures'. *Police Journal* 11 (1938): 428–49.

———. *Road Traffic and Its Control*. London: E. Arnold, 1938.

———. 'The Traffic Problem'. *Police Journal* 9 (1936): 74–97.

Watson, Henry. *Street Traffic Flow*. London: Chapman and Hall, 1933.

Wilder, Craig Steven. *A Covenant with Color: Race and Social Power in Brooklyn*. New York: Columbia University Press, 2000.

Winter, James. *London's Teeming Streets, 1830–1914*. London: Routledge, 1993.

Wootton, G. A. H. 'Mechanical Aids for Traffic Control and Other Police Purposes'. *Police Journal* 1 (1928): 89–97.

5   Economists, prices, and markets[1]

Introduction

Congestion charging was established in London in 2003, but its theoretical origins can be dated to the 1920s, and its practical development tracked back to the 1950s. It may feel like a recent introduction but its gestation took many decades. This chapter looks at its origins in road pricing, an approach to traffic originally formed by economists in direct opposition to planners' hegemonic view of hard reconstruction, and excluded by the planning profession from its dominant discourse. The intellectual and ideological trajectory of the idea of road pricing through the battleground of postwar British politics will be examined, where neoliberals influenced by the likes of Friedrich Hayek and Milton Friedman sought to counter the Keynesian consensus of centralized planning and large-state tax-and-spend. Its prime architect, Alan Walters, returned to the concept of road pricing time and again through his long career into Thatcherism and beyond.

Yet, while it was one of the earliest expressed free-market policies, road pricing did not achieve the success associated with other infrastructure market-izations such as power, water, or telecommunications, nor with wider policy areas such as health care or education. Road pricing in the neoliberal project, with its focus on detailed tracking of vehicle journeys in space and time to construct a use-based charge, foundered on traditional Tory concerns over individual freedom and big-state taxation. It also foundered on the nature of Thatcherism, which was not straightforwardly neoliberal, employing marketization but with an acute political concern with class politics and, of course, electoral success. The aspirational, lower-middle-class car driver was at the heart of these concerns. This chapter thus recounts a significant episode in traffic planning that was actively margin-alized by the planning profession but, by considering a wider network, can be seen centrally within the context of twentieth-century British political history. In this network, too, road pricing moved between the margins and the mainstream, demonstrating the portability and mutability of characterizations of traffic in socio-political contexts.

Road pricing failed as a consciously neoliberal policy, but it did not die. Instead it was transformed. Having existed since the 1960s in a parallel social democratic discourse, it was picked up in the late 1980s as an early example of the more social democratic tendencies of the Blairite turn, which used market mechanisms to achieve social ends. Here, the moral force of the argument for road pricing was transformed, with traffic congestion becoming a public bad rather than a market externality. Thus, road pricing became congestion charging. Drawing on an emerging international politics of the environment, as well as the complex local politics of London, the policy was promoted by both Conservative and Labour governments in the 1990s before being introduced (albeit in a crude form) in London by mayor Ken Livingstone in 2003 – although the policy was not his own.

How do policies travel, and how is their journey one of transformation and reinvention? This chapter will look at four key episodes. The first is the emergence of a group of transport economists, catalysed by Alan Walters, who engineered an opportunity to present their ideas on road pricing to the British government as a counterpoint to the work of urban planner Colin Buchanan. The second episode to be examined is the outcome of this work, known as the Smeed Report, and the work of its authors in negotiating a political outcome from a technical brief. The third involves the activities of British neoliberal think-tanks in promoting the ideas embodied in the Smeed Report as part of their free-market policy portfolio. They were conscious that this was a long-term project and that short-term failure might nevertheless lead in the end to success. This will take us to the final episode, in which road pricing, having failed to gain traction on the road to Thatcherism, found new meanings in social democracy amidst the environmental turn and, through changes in primary legislation enabling proceeds to be earmarked for public transport improvements, was finally implemented in London as a consciously left-wing policy.

Creating road pricing, 1920–1962

In 1951, Alan Walters, then aged 25, joined the University of Birmingham as a junior lecturer in econometrics under transport economist Gilbert Walker. One of his first major pieces of work was to develop a free-market approach to road use, using marginal cost taxation to correct the market failure of traffic congestion. 'It is clear', he wrote,

> that measures such as an annual licence fee for vehicles using the central areas of large towns would be a move in the right direction. For example, a special 'London licence' would have to be acquired and displayed before vehicles could use the roads of central London between 8 a.m. and 6 p.m.[2]

The idea of a London licence had first been proposed in 1941 by planner Stanley Adshead, who saw a return to the ancient tollgate system as the cheapest way to solve London's congestion problem, although there is no evidence Walters was aware of it.[3] While Walters was working out his ideas in the UK, the US political economist James M. Buchanan was developing ideas independently on similar lines.[4] Buchanan, who studied under Frank Knight at the University of Chicago in the late 1940s, set out to counter the prevailing doctrine in the USA of street building and reconstruction, believing that urban congestion could not be solved by engineering. Instead, he set out firm arguments for the use of 'the rationing device of the free economy, namely price', for distributing existing road space among potential users.[5]

But the work of Walters and Buchanan drew upon a theoretical expression of a market-based approach to congestion first mooted in the 1920s by two towering figures of the field, namely Arthur Pigou at Cambridge University, in a canonical 1920 work, and Frank Knight, then at the University of Iowa, in a 1924 riposte.[6] They established the theoretical problem of 'two roads'. One was good but narrow; the other wide but poor. With free access, most motorists would choose the good road, but it would soon get congested – an externality, a failure of the market. Instead, if a charge for use of the good road was levied, either by government intervention or by private enterprise, some motorists would choose to take the cheaper, poorer road, and if the charge was set just right, the good road would become just decongested enough to become free-flowing but not under-used. Market failure would be corrected and the system would be efficient, the price giving motorists the information they needed about the state of the roads ahead to enable rational decisions to be made.

But Pigou and Knight were not concerned with *real* roads. For them, traffic was a metaphor for wider aspects of political economics and market failures, and took place amidst a complex discourse within the economics discipline. As it happened, Knight won the roads battle, with Pigou quietly withdrawing his metaphor, but this barely matters. This was a time when schools of thought crystallized into teaching and research programmes, and we can see that Pigou and Knight's two roads, as well as acting as a direct influence on James Buchanan, also influenced a generation of scholars as they became embedded in a culture of economic theory that was to last for many decades.[7] One example of this disciplinary influence came from the Cowles Commission for Research in Economics. Cowles scholars took a highly theoretical and econometric approach, modelling economic systems as sets of interacting variables expressible by simultaneous equations in which, as Johan van Overtveldt has described, 'everything was linked to everything'.[8] Their Chicago base shared a building with the department run by Knight, at which economists such as Buchanan and Friedman had studied, but by the 1950s the Cowles school had developed a distinct Keynesian flavour increasingly at odds with the classical approaches taken by Knight. Relationships between the two groups frayed, and in 1955 the Commission had moved to Yale University.[9] There, Martin Beckmann and Charles McGuire completed a study of transport economics begun at Chicago with Christopher Winsten, an Oxford University economist based there from 1952 to 1954. Their research developed Pigou's two roads into an econometric model with 'efficiency tolls' levied on congested roads.[10]

The Cowles project had arisen out of work being done by economist William Vickrey on pricing structures for the congested New York rapid transit system.[11]

In 1959, following research into the technology of vehicle monitoring and toll collection, Vickrey presented evidence to the US Congress that offered the first detailed account of a practical congestion charging system.[12] Vickrey's work was technically advanced and set the foundations for much subsequent work on road pricing, drawing from vehicle tracking technologies developed for freight railways and bus companies as well as computerized pattern-recognition systems that analysed images.

But Vickrey's work also raised the spectre of the politics inherent in the scheme. His own analysis was that 'while this makes sense to economists, it seems to be politically ... somewhat unpalatable'. The committee chairman's response was blunter: 'I would not think it would be a good platform, perhaps, Professor, to launch into a political career on.'[13] Yet a highly significant political career was indeed being launched on that platform, as Alan Walters had by then begun fleshing out his earlier ideas on congestion pricing.[14] In this new work, Walters applied a recently published methodological model by Milton Friedman that combined both theoretical and empirical analysis of real-life problems using statistics and econometrics.[15]

In Britain, these strands were drawn together at the government's Road Research Laboratory (RRL), which played a crucial role in formalizing and legitimizing the political and economic concept of road-use charging. A 1960 RRL report on urban traffic congestion tolls was followed in 1961 by an influential paper by the laboratory's deputy director of traffic and safety, transport statistician Reuben Smeed.[16] In 1961, Gabriel Roth, a transport economist at Cambridge University who had worked at the laboratory from 1956 to 1959, published a paper in which he expressed exasperation that progress had not been made on this matter sooner. The case for a congestion tax, Roth observed:

> was argued by Pigou as long ago as 1920. Economists who have studied the problem agree that a tax on vehicles using congested streets is the appropriate economic answer to the problem. Why has this not been widely advocated?[17]

Yet Roth's wish was about to be granted. The early 1960s were a time of great change in British transport politics. Ernest Marples, the transport minister from 1959, instigated major studies into the railways, led by Richard Beeching, and urban traffic, led by Colin Buchanan and discussed elsewhere in this book. Yet a third study, led by Reuben Smeed, emerged from his department, which is less well known, and which drew on the distinct approach taken by the transport economists.

The Smeed Report, 1962–1964

It is important to observe from this history that road pricing was not created as a neoliberal idea, though some of its early proponents could be described as neoliberals. Keynesians, welfare economists, and technocrats were developing such ideas too, but it was to be the neoliberals who took them up most enthusiastically in the 1960s, partly because they struggled least with the distributive aspects of road pricing. This complexity can be clearly discerned in the Smeed committee's work. In May 1962, Smeed, Gabriel Roth, Alan Walters, Christopher Winsten, Michael Beesley from the London School of Economics (LSE), and Christopher Foster from Oxford were invited to join Ministry of Transport officials in a discussion on road-use charging, which took place on 12 July.[18] Following a wide-ranging exchange of views, the meeting concluded that further research was needed on the technical possibilities, with Smeed agreeing to chair a committee to carry out the research. Two years later, the group's report, entitled *Road Pricing: The Economic and Technical Possibilities*, was published, but received nothing like the attention heaped on Colin Buchanan's *Traffic in Towns*, published a few months earlier.[19] It was not that Smeed's group lacked experience or political clout – Walters and Foster both went on to play crucial roles in British political life. Yet it is *Traffic in Towns* that has come to be seen as the paradigmatic statement on British postwar urban transport policy – in opposition to the views of the economists, as both Peter Hall and Simon Gunn have observed.

Noting a 'broad consensus' that emerged on the aptness of the proposals outlined in *Traffic in Towns*, including motoring pressure groups, professional periodicals, and the government, Gunn explained as follows:

> The only dissenting voices appeared to be from the transport economists, who argued that Buchanan had overlooked the effects of urban dispersal, exaggerated the future rate of traffic growth, and ignored the possibility that environmental concerns could be solved more effectively by the market than statutory means.[20]

For Hall, this clash between Buchanan's absolute environmentalism and the relativism of the economists took the form of 'fundamental and repeated conflict'.[21]

Hall and Gunn's perceptive observations aside, there has so far been little critical attention paid to *Road Pricing* when compared with that afforded to *Traffic in Towns*. This may be due to the limited scope of Smeed's work, which was restricted to examining details of economic theory and technologies of charging. Studies of road pricing history have generally taken place internally within the disciplines of economics and transport planning and have situated the Smeed Report on a linear progression, although Martin Richards's comprehensive insider's account and Pierre-Henri Derycke's subtle comparative analysis stand out in exception.[22] Most others have taken the form of extended literature reviews.[23]

Yet *Road Pricing* was about more than just traffic – it expressed a political view and was born out of dissent with the planning approach that defined Colin Buchanan's *Traffic in Towns* study. In 1961, Roth had launched a stinging attack on congestion relief by material intervention:

We are often told that the alternatives before us are either the reconstruction of our city centres, or else slow stagnation as more and more vehicles try to use the limited road space available. Are there no other possibilities?[24]

He concluded: 'Before pulling down the fabric of the country let us consider at least the economic implications of what we are doing.'[25]

In April 1962, Roth authored a paper with Beesley, who had worked alongside Walters at Birmingham University through the 1950s, which they submitted to Buchanan's steering committee secretary.[26] It was backed by Walters, Foster, and four other eminent economists, and in their covering note they stated: 'We are convinced that little progress can be made in alleviating the difficulties caused by urban traffic congestion until the underlying economic factors are recognised and taken into account by traffic and town planners.'[27] A couple of months later they submitted a further paper.[28] Both were sent to Buchanan's steering group in a pack of papers for their July 1962 meeting, but the group's secretary discouraged any discussion of them, directing that 'Mr Buchanan hopes that the Steering group will not spend too much time on them', as they were 'not very closely related to his line of work'.[29] Buchanan himself told them that he was against the idea.[30]

Roth later recalled that Buchanan's team 'had little use for road pricing', an opinion echoed by Peter Hall, but that Foster had showed the economists' paper to Ministry of Transport officials who seemed interested in its contents.[31] The Ministry of Transport meeting in July 1962 and the Smeed Committee that followed were the result. But if the planners and the economists were at odds, it is also true that all was not well between the economists themselves. The committee's first draft report was ready in January 1963. It explained:

> we have sought to avoid the major social and political issues inherent in our subject. We are, however, fully aware of their importance, and we would not wish our approach to the subject to be misinterpreted on account of our limited terms of reference.[32]

The Ministry of Transport had forced them to stick to econometrics and technologies but, as they considered devices ranging from sensors to vouchers and meters, some on the committee had a problem.

As the draft reports were prepared and discussed, we can discern a political battle about the proper role of the panel. On one side were Walters and Roth, who saw the committee's work as an opportunity for a wider political exercise. Their proposal was that road pricing should be partnered with reductions or even abolition of the two main charges motorists already incurred, namely the annual licence fee and fuel duty. This was an argument for a full overhaul of the motor tax system, with revenues from road pricing earmarked for road spending. Their goal was no net increase in revenue raised under the new road pricing system but rather a redistribution of costs according to the market mechanism. Winsten and the Ministry officials felt, however, that this was overstepping the panel's remit by making political recommendations masquerading as economic theory.

Winsten felt the first report draft was 'still far too slanted towards particular points of view and far too little a technical assessment of different methods of road charging'. Raising the spectre of income tax redistribution if the charging scheme was part of a wider taxation overhaul, as the draft suggested, he observed that 'In discussing these questions one is essentially striking a political attitude and this is inappropriate for the Committee to do, nor should it conceal a political attitude under an apparently technical discussion.' He concluded:

> I feel that the report tries much too hard to advocate one particular solution rather than listing possibilities ... If it were put through in its present state then there would be many parts to which I would feel very reluctant yet to put my signature.[33]

In his own response, Ministry economist Richard Bird (later Barbara Castle's Principal Private Secretary) said 'this whole passage reads, in my opinion, rather like a party political speech', which contained a 'considerable number of unsupported value judgment[s]'.[34]

Roth felt that an opportunity to press for a wholesale motoring tax overhaul was slipping from his grasp, owing to cautious officials and academic economists debating the niceties of theory. He wrote an exasperated letter to his ally, Walters, following the interventions of Winsten and Bird. Decrying the 'nebulous and unconstructive approach of most of the economists', he exclaimed 'Here we have a Government coming to us and wanting to know what they should do about the pricing of roads, and all we seem to say is that the problem raises great difficulties in the economic field.' He concluded by saying 'If we cannot explain to a layman the things on which we are agreed, how on earth can we expect him to provide the funds to enable us to explore the things that need exploring?'[35]

To circumvent the political veto imposed on their work, Roth led a publishing campaign, sending one of the papers they had originally submitted to the Buchanan Committee onwards to the *Town Planning Review*.[36] He also wrote about income distribution in *Crossbow*, the magazine of the Conservative Party's Bow Group, and co-authored a paper with committee secretary Michael Thomson for the monthly magazine *Aspect* on road pricing as part of a wider road tax overhaul.[37] Foster, too, had been writing on 'the transport problem', publishing a book in 1963 that rather tersely observed that the Smeed panel report was 'as yet ... unpublished'.[38] In fact the committee's final report had been submitted to the Ministry of Transport in June 1963. It concluded that 'there is every possibility that at least one of these proposals could be developed into an efficient charging system and could yield substantial benefits on congested roads'.[39] Yet the report was destined to sit on Transport Minister Ernest Marples's desk for a full year before it was released, overtaken by the publication of Buchanan's *Traffic in Towns* in November 1963.

For Marples and for the Treasury, the Smeed Committee's report could not have been less welcome. It flew in the face of two political truisms. Firstly, motorists resented being taxed. Secondly, the Treasury did not believe in taxes being specifically earmarked – or 'hypothecated' – to any particular category of spend.[40] It is notable that the Smeed panel had not been instigated by Marples but by his officials, and the Minister had nothing to do with it until its report was completed.[41] In fact, in a recent memoir, Foster has suggested that Marples tried to stop the report's publication but was unable to do so directly because the RRL was part of the Department of Scientific and Industrial Research, not his own Ministry of Transport.[42] Furthermore, the Ministry's hand had been forced by Roth's publication drive. In September, Marples agreed to publish, but only with major revisions.[43] Winsten again expressed his concerns and later revealed tension in the committee's meetings, describing discussions as 'harrowing' and observing that some of his reservations had not been discussed, 'given the time and atmosphere'.[44]

Roth was appalled at the delay, writing to the Ministry on 15 November to complain that no publicity was planned for their work, and also the following:

> no effort is being made to relate the publication time of our Report to the publication time of the Buchanan Report ... You now have the two reports. They are very relevant one to the other. It would surely be better for them to appear as part of a coherent pattern than as a couple of unconnected items.[45]

He had previously expressed grave concern about the nature of publicity surrounding the report, having told Smeed the following:

> When all is said and done, very few people will read our Report; most will rely on the newspaper reports which will depend on the press hand-outs. I feel that it is essential for the RRL to have a hand in preparing these, and I would gladly be of assistance in this respect. We must avoid being stabbed in the back by a hostile or ignorant press officer.[46]

But it was not until 25 November that Winsten finally gave his approval to the report's wording, and by then it was too late to publish alongside the Buchanan Report, which was released two days later.[47] Perhaps this had been the Ministry's intention, but in any case, the report was still in the sand, as the Treasury still objected to hypothecated tax revenues.[48]

Roth had by now despaired of the report being published with any meat on its bones. Responding to Smeed's proposals for further amendments, he complained of the 'Treasury's noose round our necks' and pushed for the panel to publish its original report by itself, stating the following:

> The position of the Ministry and RRL members of the Panel is clearly somewhat delicate, but other members of the Panel are not so constrained. We can command some influential support, even within the Treasury. Why

should we put up with this shabby treatment? If we wait much longer the election may be on us.[49]

Roth was right to be concerned; it has been said that Alec Douglas Home, who replaced Harold Macmillan as Prime Minister in October 1963, said to officials 'let us take a vow that if we are re-elected we will never again set up a study like this one'.[50] By March, though, the report had been watered down enough to secure Treasury approval, and Marples told the House that it was to be published.[51] But Roth was, in the end, unable to garner the publicity fanfare that had accompanied Buchanan's ideas, or to retain in the final Smeed Report the wider issues of taxation and distribution that had been proscribed from the start. While Marples finally released the report in June 1964, he did so with the minimum of fanfare.[52]

The Smeed Report had finally joined the Buchanan Report in public view, but the two sides had not reached any kind of rapprochement, as was witnessed first-hand by Peter Hall. On Buchanan's side, dealing with environmental concerns such as noise and congestion was an absolute requirement. The economists, Hall observed, 'had a hard time with that, because it flatly contradicted their basic assumption that everything had its price and that everything was tradable off against everything else'. Hall recalled attending a lecture at the LSE in 1964 'when Buchanan confronted Beesley and Foster; there was no meeting of minds at all, just mutual incomprehension'.[53] But outside official political channels, road pricing gained new life.

Think-tanks and the neoliberal project, 1963–1989

Four months after the Smeed Report's publication, Labour won the general election and Harold Wilson took over the premiership. British politics was still in the postwar consensus where Keynesian policies formed an orthodoxy across the mainstream parties. But Daniel Stedman Jones, Richard Cockett, and others have described the development of a concerted neoliberal project in the mid-twentieth century dedicated to fighting the consensus, where the term 'neoliberal' is defined by Stedman Jones as 'the free market ideology based on individual liberty and limited government that connected human freedom to the actions of the rational, self-interested actor in the competitive marketplace'.[54] This project involved the development of ideas expressing the neoliberal viewpoint to a broad range of opinion-formers – academics, students, journalists, and politicians – in the hope, over a 20-year period, of reversing the postwar consensus.

Here, economics and politics were tightly intertwined, and the Buchanan–Smeed episode represents these polar opposites in political economy. Colin Buchanan and the postwar urban planners represented the consensus, with *Traffic in Towns* enshrining the centrally planned city built through Keynesian ideology on the proceeds of taxation or deficit spending. Reuben Smeed, on the other hand, led a discourse that could represent individualism and economic liberalism, with *Road Pricing* expressing the marketization of road space paid for by those who use it. The work of Alan Walters in the 1950s and on the Smeed Committee is notable, given his later prominent role in the Thatcherite neoliberal revolution. By the mid-1960s he was, as Cockett observes, 'an ardent economic liberal'.[55] By 1975, having advised and been sacked by Edward Heath, he was informally advising Margaret Thatcher on her economic approach to government based on a broadly neoliberal agenda, and in 1981 she hired him as her personal economic adviser to steer through an austerity budget. In 1989 he returned to her side – leading to the resignation of her Chancellor, Nigel Lawson.[56] Through all this, he continued to work on road pricing.[57] The origin of congestion pricing proposals in the work of public-choice theorist James Buchanan is equally suggestive of a neoliberal reading, with Stedman Jones placing him alongside other US thinkers such as Milton Friedman and George Stigler as 'intellectual nodes at the heart of a transatlantic network ... who spread an increasingly honed political message of the superiority of free markets'.[58]

Within that network, neoliberal think-tanks played a highly significant role. At a time when the political consensus meant that discussion of anti-Keynesian policies fell largely on deaf ears across the political spectrum, it fell to outsiders to debate, develop, and promote economic liberalism. Until the mid-1970s, one institution held the reins in the UK: the Institute of Economic Affairs (IEA), described by Stedman Jones as 'the single most important organization promoting neoliberal ideas in the 1960s'.[59] Founded in 1955, the organization's purpose was to develop a long-term programme of research and dissemination of market-economy policy ideas to secure a 'transformation of the intellectual climate', as Cockett has put it.[60] Simplification of the neoliberal message for public consumption was crucial.[61]

From the outset, the IEA therefore exploited the power of publication, and appointed economist and writer Arthur Seldon as its editorial director. Its publications were written by well-informed academic economists and were short, pithy, and cheap. The point was to build up a wide range of ideas, each promoting the market economy, that together could form an intellectual manifesto for change. As part of his programme, Seldon commissioned a series of publications on the financing of industry and social services. Alan Walters and James Buchanan both wrote for the series.[62] With the disappointing reception of the Smeed Report in 1964 fresh in the minds of Walters and others, Seldon saw urban traffic as a useful policy area for promoting market ideas, and commissioned a total of three books on the topic. Gabriel Roth wrote the first two. His first, *Paying for Parking*, was published in 1965. In Seldon's preface, he observed that 'Roads must in practice be rationed in one way or another: by edict, licence or congestion – or by price.' He went on: 'In distinction to the solutions usually offered by physical planners and social engineers, Mr Roth has proposed a structure of market prices.'[63] The following year, Seldon published Roth's *A Self-Financing Road System*, in which Roth outlined his Smeed Committee plans for a total overhaul of the tax system for motorists in favour of a system based entirely on road-use pricing.[64]

But Seldon also wanted a direct neoliberal critique of Colin Buchanan's *Traffic in Towns*, so in 1966 he turned to transport economist Dennis Reynolds, who had worked in the late 1950s with Michael Beesley at the RRL on the cost–benefit analysis of the M1 motorway.[65] In 1963, having left the laboratory, Reynolds had published an article expressing disappointment that 'planners seem to have taken little interest in economics (or, with certain exceptions, economists in planning)', painting a picture of idealistic, other-worldly planners who would benefit from a dose of reality:

> not only might the framework and methodology of economics, e.g. divergence between private and social costs, be of general help to the planner in his daily task, but it might help to reconcile him to the logic behind many of the harder realities of economic life, e.g. the scarcity of resources.[66]

Reynolds's 1966 IEA monograph focused this rhetoric in a direct attack on Colin Buchanan. In it, he sought to 'call a timely halt to the divorce between the technical proposals of enthusiasts and economic realities'. Described by Seldon as 'originally a Buchanan enthusiast', Reynolds dismissed planning policy as arbitrary, in Seldon's words, and expressed doubt, following recent work by James Buchanan, that the state could ever direct resources efficiently.[67] But his target was not just the planner; he also criticized cost–benefit analysis being championed by Beesley, seeking to distance himself from their joint work carried out in the 1950s.[68] Ultimately, his proposals took the form of population dispersal to new towns with an overlay of market pricing to curb the excesses of the urban planners. In this mode, congestion pricing was a useful tool in rendering busy urban centres less attractive, with the anticipation that flight to the suburbs would follow.

In the end, Reynolds's work simply pointed up the inability of the economics discipline to present a united message – as Roth had observed many times – and appeared to the wider world as an austere and theoretical opposition to the potential revolution called for by Colin Buchanan. The book was reviewed separately by Beesley and Winsten, both of whom were decidedly lukewarm in their assessment.[69] It was then left to transport engineer Dennis Gilbert, who worked with Buchanan at his newly formed consultancy, to deliver the final blows. He found Reynolds's assessment of the costs of Buchanan 'rather misleading as well as lacking rigour', and considered that 'It seems clear that the author has not troubled to check his ideas about the report by careful reading.'[70]

We have so far charted a subtle transformation of road pricing from its inception in 1920s welfare economics to 1950s and 1960s counter-Keynesian economic liberalism, even if its early promoters such as William Vickrey were not involved in the neoliberal project. But, while other market policies, such as those relating to health care and education, took root in Margaret Thatcher's government after 1979, road pricing did not, as we will see. Yet, for Seldon, Reynolds's book served its purpose in contributing to the intellectual debate on resource allocation and raising the idea of markets for road space. It was not unusual for individual neoliberal ideas to fail, according to one prominent free-marketeer in 1968:

> It is scarcely an exaggeration to say that liberalism claims to cure more ills than patent medicine. Are you worried about the world monetary crisis? Flexible exchange rates will ease your mind. Is urban congestion a problem in your town? Enquire about road pricing today. Have you lain awake at nights, praying for choice in education? The voucher scheme is meant for you. Do high prices make you feel faint? Abolition of RPM will soon do the trick. Naturally it is easy to mock at this out-pouring of ideas – *and probably some of them will not work* – but it is the mark of a dynamic philosophy that it provides clear, consistent and above all, simple solutions to the problems thrown up by society and the economy.[71]

What mattered most was the development of a wider argument applicable in any receptive political environment, and in 2003 road pricing was introduced – under a Labour government.

The environment, social democracy, and the politics of London, 1963–2003

Road pricing was a neoliberal fire that burnt brightly and then burnt out – as many did. But the promotion of road pricing in the neoliberal think-tanks through the 1960s and beyond took place against a broader political background, and road pricing was also considered as part of a broad social democratic discourse, which embraced it *within* the planning paradigm rather than in opposition (where the term 'social democracy' is defined by the *Oxford English Dictionary* as 'a system of liberal democracy that retains a commitment to social reform and social justice').[72] Over the year 1962–3, in parallel with the sitting of the Smeed Committee, the journal *Socialist Commentary* convened a group of transport specialists to discuss transport from a 'left' perspective. As Joe Moran has observed, *Socialist Commentary* had long advocated the targeting of floating voters during Labour's long period in opposition, and in *Transport Is Everyone's Problem*, the pamphlet written by group member Peter Hall that emerged from the meetings, a direct appeal was made to the 'memory of Orpington', claiming that 'the anger of commuters in a hundred such constituencies may decide the outcome of the next General Election'.[73]

The group strongly advocated Smeed-style road pricing in urban areas as a 'first essential for intelligent city traffic planning' and, as the Smeed Committee was discussing at the same time, also advocated a taxation overhaul, stating the following:

> The introduction of 'congestion charging' would not mean an increase in the total level of motor taxation, merely a rearrangement of the burden ... The

object is merely to relate the burden of motor taxation more accurately to the true costs of each journey (including the social costs) than is attempted at present.[74]

Yet in this context, road pricing played just one part in a total transportation plan, which recognized that 'transport planning is a central part of positive national and regional, economic and physical planning'.[75]

The common denominator across both the *Socialist Commentary* and the Smeed groups was the economist and civil servant Christopher Foster, who acted as a conduit for ideas to pass from the latter to the former. Foster, who later advised on rail privatization and the poll tax, but was also an adviser to Anthony Crosland and Barbara Castle, cannot be situated on a simple political spectrum. This episode therefore undermines the notion of an unproblematic, fixed consensus in British politics in the 1950s and 1960s. As Daniel Stedman Jones has observed, the history of market-based politics at this time was one in which 'ideas move, change, become distorted, and sometimes even mix with their polar opposites in the messy world of electoral politics, policy and government'.[76]

Whether neoliberal or social democratic, road pricing was not taken up in the 1960s.[77] Labour's October 1964 election win signalled a period of ambivalence towards the policy. Ernest Marples was replaced first by Tom Fraser and then, from December 1965, by Barbara Castle. Both ministers took steps to continue the analysis of road pricing, but with little zeal. In May 1965, Fraser appointed a body to develop the work of both the Buchanan and Smeed groups, looking at methods of restraint more broadly than just by price.[78] Castle made clear her own support for congestion pricing, countering suggestions that such a policy would favour the rich.[79] But it was clear that this was not a popular scheme. Both the Conservative Party and a range of road user groups, including the Automobile Association (AA), Royal Automobile Club (RAC), and British Roads Federation, lobbied hard against the ideas, and the result was a 1967 report that concluded that short-term restraint would be most easily achieved by parking controls in central London, with Smeed-style road pricing put on the back burner.[80] Smeed Committee secretary J. Michael Thomson responded with details of a simple daily licence scheme for central London, and further experiments were carried out on road-pricing meter technology, but neither licences nor meters came to anything in 1960s London. A further powerful influence against any restriction on car use came from the British car manufacturing industry, as Simon Gunn has argued.[81]

Into the 1970s, under Edward Heath and Margaret Thatcher, there was little appetite to revive the scheme, despite Walters's prominent role in setting economic policy. The prevailing dogma, 'kept alive by the AA and the RAC', as William Plowden put it, was twofold: that motorists possessed the inviolable freedom to drive without restriction, and that motor taxation was too high and its proceeds not correctly applied to the road system.[82] Despite dreams of a total overhaul of motor taxation, in reality the market principle of road pricing appeared irreconcilably to signal both an increase in taxation and a loss of freedom, unacceptable in Thatcher's car-owning democracy.[83] Through the 1970s, parking restraint formed the principal focus for Tory urban congestion-management policy.[84]

A form of road pricing was eventually introduced in London by the left-wing mayor Ken Livingstone in 2003, but the political shifts that enabled it began in the 1980s.[85] Treasury policy changed to permit the proceeds of a congestion charge to be hypothecated to public transport improvements, owing in large part to a major shift in global politics towards environmentalism some 25 years after Colin Buchanan raised the flag. At the national level, this saw transport policy start to move from a 'predict and provide' model to a more contested approach focused on a 'new realism'.[86] Concern over the environmental effects of transportation had been growing since the early 1960s. In 1987, the World Commission on Environment and Development published a highly influential report, which catalysed policy debates over the environment.[87] In 1989, Department of Transport predictions of a huge growth in car use led to the publication of a multi-billion-pound road building programme, but the predictions also prompted concerns about the effects of unchecked traffic and the inability of road building alone to solve congestion.[88] The politics of transport were becoming more contested and, after the shock 1989 growth forecasts were digested, new approaches on an environmentally aware restraint model were able to gain space at the table – with road pricing linked to wider packages of public transport improvements in London and other urban centres.

A slew of reports was published from 1989 to 1991, recognizing the need for balanced transport policies with some element of road pricing.[89] One was particularly significant, for while neoliberal think-tanks were still occasionally promoting the policy from a right-wing perspective (with Gabriel Roth continuing to lead the charge), Patricia Hewitt, deputy director of the newly formed Institute for Public Policy Research (IPPR), offered a consciously left-wing interpretation in the group's first Green Paper, published in 1989:

> On the political right (in, for instance the proposals of the Adam Smith Institute) it is seen as an extension of free market principles ... On the left, road pricing is often condemned because it appears to favour the rich at the expense of the poor and average citizen. IPPR takes a different approach ... Road pricing – or, more accurately, charges for car use on heavily congested roads – are an important application of the 'polluter pays principle', a principle to which lip-service at least is paid by conservative, liberal and left politicians alike.[90]

That both the Adam Smith Institute (ASI) and the IPPR were promoting free-market approaches to roads lends weight to Richard Cockett's analysis that the new left consciously modelled its policy-making strategy on the neoliberal think-tanks such as the IEA and its younger siblings, the ASI and the Centre for Policy Studies (CPS).[91] Clearly, policies as well as strategies were borrowed and transformed by all sides. One ASI proposal, written in 1989, proposed road pricing with hypothecation to the urban bus network – but a deregulated, free-market bus network – and it was a 1995 CPS paper that described most closely the technological approach adopted in 2003, where the congestion charge was enforced by automatic number-plate recognition cameras.[92] But Hewitt's rhetorical transformation of the argument from road pricing (as it tended to be termed in the neoliberal discourse) to congestion charging (which incidentally had been Hall's terminology in the 1963 *Socialist Commentary* report) signalled a shift away from the motor taxation debate towards the marketization of environmental pollution.

The journey taken by road pricing from Hewitt's crucial 1989 IPPR paper to the 2003 London congestion charge is described comprehensively by Martin Richards.[93] To summarize, this was a combination of Labour and Conservative policy. Hewitt's argument persuaded John Prescott to include road pricing, with proceeds hypothecated to public transport improvements, in his 1989 transport policy statement.[94] In 1991, the Department of Transport set up a London congestion charging study, later sponsored by the Government Office for London (GOL) following its creation under John Major's government in 1994, which reported in July 1995 in favour of a London congestion charge.[95] Meanwhile, in 1992 the UN Earth Summit had been held in Rio de Janeiro, focusing particular attention on the environmental impact of road transport, and attention in Britain had reached a critical moment in 1994 with the publication of a Royal Commission on Environmental Pollution report, which was highly critical of the government's transport policy, recommending a reduction in road traffic growth and expansion of public transport.[96]

The GOL report was kicked into touch by the Conservative government.[97] However, the incoming Labour government seized the opportunity to respond to the environmental debate by rolling together the ministerial responsibilities for transport and the environment into a single department under the deputy Prime Minister, John Prescott, who, as we have seen, was already in favour of a congestion charge if it enhanced public transport.[98] In July 1998, he published a White Paper announcing local authority powers to charge for road use.[99] At that stage, work was in progress, led by the GOL and its director, Genie Turton, in drawing up legislation to establish the Greater London Authority, led by an elected mayor. A new research project was formed, led by Turton, looking at powers for the new mayor to introduce road charging, which reported in March 2000.[100] In July, the winner of the mayoral elections, Ken Livingstone, took up his post and immediately began setting up a congestion charging scheme.[101] It is important to remember that the London congestion charge was not Livingstone's own policy – it was an option available to whichever candidate won the mayoral election, and it had been long in development – but its conversion from policy to reality was no easy task. Livingstone's opportunism and his grasp of public politics finally broke the logjam. After much lobbying of stakeholders, a workable – and, moreover, politically palatable – charge scheme was created. Livingstone's rhetoric – framed around the damage caused by congestion to the environment and people's quality of life – echoed that of Patricia Hewitt over a decade earlier.[102] In February 2003, the London congestion charge went live. Described by Tony Travers as 'the single most radical policy introduced by Ken Livingstone in his post-2000 administration', its subsequent history is a story only now being told.[103]

Conclusions

Debates over traffic congestion mirrored wider concerns over the politics of planning. In the post-war Keynesian consensus, it made sense that centralized urban planning was the dominant discourse in which congestion was framed, but this chapter has described an alternative, marginalized view. At its most straight-forward level this was, as Peter Hall has argued, a disagreement over standards – in this case environmental ones – with Colin Buchanan considering them absolute and the economists believing they were relative and thus tradeable. Looking broadly, however, this appeared to map onto a world view that abhorred state-controlled allocation of resources in principle, and for which Buchanan's proposals for expensive state-funded interventions in the urban realm were anathema. Thus, we see the agitation for an alternative committee – Smeed – which attempted at least to temper the perceived excesses of the town planners and vest some control in individual drivers, and we see the active participation of the IEA and Alan Walters in promoting a policy of road pricing in the neoliberal context through specialist transport economists such as Gabriel Roth and Dennis Reynolds. *Traffic in Towns* and *Road Pricing* represented far more than alternative solutions to congestion. They expressed a fundamental and critical debate in mid-twentieth-century British history.

For the IEA, congestion charging was just one more free-market policy idea, and the social and cultural context in which an idea is floated makes a difference to its reception, as do local politics, which vary hugely in time and place. The success or failure of congestion charging as a policy was not, therefore, inevitable. As it happens, it did not catch on during Margaret Thatcher's reign, as other marketizations did. There was, of course, more to Thatcher than neoliberalism, and her pragmatic class politics, framed around the aspirational car driver and a long memory of Orpington man, meant road pricing was an unlikely policy for a leader famously dismissive of public transport. More fundamentally, road pricing clashed with the great Tory beliefs in individual freedom and low taxation, as without a wholesale prior disman-tling of the existing motor taxation system – which the Treasury was not prepared to entertain – road pricing looked like additional expense for drivers, and any system that tracked the movement of citizens going about their daily business looked decidedly like a loss of liberty. As William Vickrey had said in 1959, it was always destined to be 'politically unpalatable' when framed in an economic discourse. It is perhaps not surprising, then, that road pricing became one of the policy ideas that failed to gain traction in the 1960s and 1970s.

But transport is an area of politics that always evades simple categorizations, presenting a Janus face and reflecting multiple, competing concerns that can only be teased out by considering a wide network of actors, ideas, technologies, and politics. These multiple concerns were expressed in 1963 in the *Socialist Commentary* report arguing for both planning and marketization of road space, and in the New Labour world of Tony Blair's government, the manifestation of urban road pricing that finally emerged from over 80 years of discussion could be seen as social democratic or neoliberal; right-wing or left-wing; an expression of markets, or of state planning, or of all these things. It was the moral dimension of the argument that had changed, in which congestion had been reframed from a market failure to an environmental evil, and free-flowing traffic framed as a social as well as an economic end. Whatever the origins of road pricing, or the journey it took, it was neither inevitable nor fixed in meaning. With a global focus on the environment, and local shifts in the politics of the markets and of London, a context emerged in the 1980s and 1990s that saw value in restricting car use and hypothecating charge revenues to public transport. This was the scheme Ken Livingstone introduced in 2003, but only because of that 'messy world of electoral politics, policy and government'.

Notes

1. A slightly modified version of this chapter was published as David Rooney, 'The Political Economy of Congestion: Road Pricing and the Neoliberal Project, 1952–2003', *Twentieth Century British History* 25, no. 4 (December 2014): 628–50, and is reproduced here by permission of Oxford University Press.
2. Alan Walters, 'Track Costs and Motor Taxation', *Journal of Industrial Economics* 2, no. 2 (1 April 1954): 143; Walters expands on his 1950s transport work in Alan Walters, 'A Life Philosophy', *American Economist* 33, no. 2 (1 October 1989): 18–24.
3. Stanley Adshead, *A New England: Planning for the Future* (London: Frederick Muller, 1941), 126–9.
4. For Walters's assertion that his ideas were developed independently, see Ministry of Transport, *Road Pricing: The Economic and Technical Possibilities* (London: HMSO, 1964), 14.
5. James M. Buchanan, 'The Pricing of Highway Services', *National Tax Journal* 5, no. 2 (June 1952): 98.
6. Arthur C. Pigou, *The Economics of Welfare* (London: Macmillan, 1920); Frank H. Knight, 'Some Fallacies in the Interpretation of Social Cost', *Quarterly Journal of Economics* 38, no. 4 (August 1924): 582–606.
7. See for instance Howard Ellis and William Fellner, 'External Economies and Diseconomies', *American Economic Review* 33, no. 3 (September 1943): 492–511.
8. Johan van Overtveldt, *The Chicago School: How the University of Chicago Assembled the Thinkers who Revolutionized Economics and Business* (Chicago: Agate, 2007), 37.
9. Carl F. Christ, 'The Cowles Commission's Contributions to Econometrics at Chicago, 1939–1955', *Journal of Economic Literature* 32, no. 1 (1 March 1994): 35; Overtveldt, 37–8; see also Jamie Peck, *Constructions of Neoliberal Reason* (Oxford: Oxford University Press, 2010), Chapter 3.
10. Martin J. Beckmann, Charles Bartlett McGuire, and Christopher Winsten, *Studies in the Economics of Transportation* (New Haven: Yale University Press, 1956).

11 William Vickrey, *The Revision of the Rapid Transit Fare Structure of the City of New York*, Technical Monograph 3 (New York: Mayor's Committee on Management Survey of the City of New York, 1952); on Vickrey's influence on the Cowles team, see Winsten's comments in J. Michael Thomson, 'An Evaluation of Two Proposals for Traffic Restraint in Central London', *Journal of the Royal Statistical Society Series A (General)* 130, no. 3 (1967): 367.

12 William Vickrey, 'Evidence', in *Transportation Plan for the National Capital Region: Hearings before the Joint Committee on Washington Metropolitan Problems (86th Congress)*, ed. US Congress (Washington: US Congress, 1959), 454–585; extracts later published as William Vickrey, 'Statement to the Joint Committee on Washington, DC, Metropolitan Problems', *Journal of Urban Economics* 36 (1994): 42–65.

13 Vickrey, 'Evidence', 464–5.

14 Alan Walters, *The Theory and Measurement of Private and Social Cost of Highway Congestion (Discussion Paper)*, Faculty of Commerce and Social Science Discussion Papers B1 (Birmingham: University of Birmingham, 1959); Alan Walters, 'The Theory and Measurement of Private and Social Cost of Highway Congestion', *Econometrica* 29, no. 4 (1 October 1961): 676–99; Alan Walters, 'Empirical Evidence on Optimum Motor Taxes for the United Kingdom', *Journal of the Royal Statistical Society, Series C (Applied Statistics)* 10, no. 3 (1 November 1961): 157–69.

15 Walters, 'A Life Philosophy', 21; for Friedman's model see Milton Friedman, *A Theory of the Consumption Function* (Princeton: Princeton University Press, 1957).

16 J. C. Tanner, *A Theoretical Study of the Possible Use of Tolls to Relieve Traffic Congestion in Urban Areas* (Road Research Laboratory RN/3819/JCT, 1960); Reuben Smeed, *The Traffic Problem in Towns* (Manchester: Manchester Statistical Society, 1961).

17 Gabriel J. Roth, 'A Pricing Policy for Road Space in Town Centres', *Journal of the Town Planning Institute*, November 1961, 287.

18 For the full committee membership, see Ministry of Transport, *Road Pricing*, iv.

19 Ministry of Transport, *Road Pricing*; Ministry of Transport, *Traffic in Towns: A Study of the Long Term Problems of Traffic in Urban Areas* (London: HMSO, 1963).

20 Simon Gunn, 'The Buchanan Report, Environment and the Problem of Traffic in 1960s Britain', *Twentieth Century British History* 22, no. 4 (2011): 532.

21 Peter Hall, 'The Buchanan Report: 40 Years On', *Proceedings of the Institution of Civil Engineers: Transport* 157, no. 1 (February 2004): 7.

22 Martin G. Richards, *Congestion Charging in London: The Policy and the Politics* (Basingstoke: Palgrave Macmillan, 2004); Pierre-Henri Derycke, 'Road Pricing: Analysis and Policies (a Historical Perspective)', *Recherches économiques de Louvain* 64, no. 1 (1998): 63–74.

23 Robin Lindsey, 'Do Economists Reach a Conclusion on Road Pricing? The Intellectual History of an Idea', *Econ Journal Watch* 3, no. 2 (May 2006): 292–379; Kenneth Button, 'The Rationale for Road Pricing: Standard Theory and Latest Advances', *Research in Transportation Economics* 9 (2004): 3–25; David Starkie, *The Motorway Age: Road and Traffic Policies in Post-War Britain* (Oxford: Pergamon Press, 1982); Andrew Evans, 'Road Congestion Pricing: When Is It a Good Policy?', *Journal of Transport Economics and Policy* 26, no. 3 (September 1992): 213–43; Steven Morrison, 'A Survey of Road Pricing', *Transportation Research A* 20, no. 2 (1986): 87–97.

24 Roth, 'A Pricing Policy', 287.

25 Roth, 'A Pricing Policy', 289.

26 The National Archives (hereafter TNA), MT 128/99, 'Restraint of Traffic in Congested Areas', Gabriel Roth and Michael Beesley, 10 April 1962.

27 TNA, MT 128/99, letter, Foster *et al.* to Crowther Committee secretary, 12 April 1962.

28 TNA, MT 128/99, 'The Pricing of Road Space in Congested Areas', Gabriel Roth and Michael Beesley, 22 June 1962.

29 TNA, MT 128/99, minute, J. A. L. Gunn to *Traffic in Towns* Steering Group, 2 July 1962.

30 See Gunn, 526.

31 Gabriel J. Roth, *Roads in a Market Economy* (Aldershot: Avebury Technical, 1996), xviii; Hall, 'The Buchanan Report', 8.

32 TNA, DSIR 12/344, 'Panel on Road Pricing: Draft Report (PRP37)', January 1963.

33 TNA, DSIR 28/521, letter, Winsten to Thomson, 16 January 1963.

34 TNA, DSIR 28/521, letter, Bird to Smeed, 14 February 1963.

35 TNA, DSIR 28/521, letter, Roth to Walters, 20 February 1963.

36 Michael E. Beesley and Gabriel J. Roth, 'Restraint of Traffic in Congested Areas', *Town Planning Review* 33, no. 3 (1 October 1962): 184–96.

37 Gabriel J. Roth, 'An End to Traffic Jams', *Crossbow* (September 1962), 13–18; Gabriel J. Roth and J. Michael Thomson, 'Road Pricing: A Cure for Congestion?', *Aspect* 3 (1963): 7–14.

38 Christopher Foster, *The Transport Problem* (London: Blackie and Son, 1963), 248.

39 Ministry of Transport, *Road Pricing*, 42.

40 David Newbery and Georgina Santos, 'Road Taxes, Road User Charges, and Earmarking', *Fiscal Studies* 20, no. 2 (1999): 103–32.

41 Starkie, 43.

42 Christopher Foster, 'Michael Beesley and Cost Benefit Analysis', *Journal of Transport Economics and Policy* 35, no. 1 (1 January 2001): 11.

43 TNA, DSIR 28/521, letter, Bird to Thomson, 20 September 1963.

44 TNA, DSIR 28/521, letter, Winsten to Smeed, 25 November 1963.

45 TNA, DSIR 28/521, letter, Roth to Mills (Ministry of Transport), 15 November 1963.

46 TNA, DSIR 28/522, letter, Roth to Smeed, 14 October 1963; on the Buchanan Report publicity, see Gunn, 527, 531–3.

47 *Hansard*, HC Deb, 27 November 1963, Vol. 685, cols. 281–8.

48 TNA, DSIR 28/522, letter, Smeed to Road Pricing Panel, n.d., but 3 January 1964.

49 TNA, DSIR 28/522, letter, Roth to Smeed, 5 January 1964.

50 Comment reported by an unnamed civil servant to be contained in Ministry of Transport files, recalled in Phil Goodwin, 'Solving Congestion: Inaugural Lecture for the Professorship of Transport Policy, University College London', 23 October 1997, eprints.ucl.ac.uk/1244/1/2004_22.pdf.

51 *Hansard*, HC Deb, 11 March 1964, Vol. 691, cols. 56–57W.

52 *Hansard*, HC Deb, 10 June 1964, Vol. 696, cols. 444–50.

53 Hall, 'The Buchanan Report', 8.

54 Daniel Stedman Jones, *Masters of the Universe: Hayek, Friedman, and the Birth of Neoliberal Politics* (Princeton: Princeton University Press, 2012), 2; Richard Cockett, *Thinking the Unthinkable: Think-Tanks and the Economic Counter-Revolution, 1931–83* (London: HarperCollins, 1994); Keith Tribe, 'Britain, Economics in (20th Century)', in *New Palgrave Dictionary of Economics*, 2nd edn (Basingstoke: Palgrave Macmillan, 2008); Keith Tribe, 'Liberalism and Neoliberalism in Britain, 1930–1980', in *The Road from Mont Pèlerin: The Making of the Neoliberal Thought Collective*, ed. Philip Mirowski and Dieter Plehwe (Cambridge, MA: Harvard University Press, 2009), 68–97; Peck; Ben Jackson, 'The Think-Tank Archipelago: Thatcherism and Neo-Liberalism', in *Making Thatcher's Britain*, ed. Ben Jackson and Robert Saunders (Cambridge: Cambridge University Press, 2012), 43–61. Peck discusses definitions of neoliberalism at length in Chapter 1.

55 Cockett, 153.

56 David Smith, 'Walters, Sir Alan Arthur (1926–2009)', *Oxford Dictionary of National Biography*, January 2013.

57 Alan Walters, *The Economics of Road User Charges* (Washington, DC: International Bank for Reconstruction and Development, 1968); Walters, 'A Life Philosophy', 20.

58 Stedman Jones, 4–5.

59 Stedman Jones, 162; see also Christopher Muller, 'The Institute of Economic Affairs: Undermining the Post-War Consensus', *Contemporary British History* 10, no. 1 (1996): 88–110.

60 Cockett, 141.

61 Stedman Jones, 9.

62 Alan Walters, *Integration in Freight Transport* (London: Institute of Economic Affairs, 1968); James M. Buchanan, *The Inconsistencies of the National Health Service* (London: Institute of Economic Affairs, 1965).

63 Editor's Preface, in Gabriel J. Roth, *Paying for Parking* (London: Institute of Economic Affairs, 1965), 2, 4.

64 Gabriel J. Roth, *A Self-Financing Road System* (London: Institute of Economic Affairs, 1966).

65 T. Coburn, Michael Beesley, and Dennis Reynolds, *The London–Birmingham Motorway: Traffic and Economics*, RRL Technical Paper 46 (Harmondsworth: Road Research Laboratory, 1960).

66 Dennis Reynolds, 'Congestion', *Journal of Industrial Economics* 11, no. 2 (April 1963): 138.

67 Dennis Reynolds, *Economics, Town Planning and Traffic* (London: Institute of Economic Affairs, 1966), preface. James Buchanan's work is referenced on 10.

68 Reynolds, *Economics, Town Planning and Traffic*, 13–14.

69 Michael E. Beesley, review of D. J. Reynolds, *Economics, Town Planning and Traffic*, *Economica* 34, no. 136 (November 1967): 451; Christopher Winsten, review of D. J. Reynolds, *Economics, Town Planning and Traffic*, *Journal of Transport Economics and Policy* 1, no. 1 (January 1967): 118–19.

70 Dennis Gilbert, review of D. J. Reynolds, *Economics, Town Planning and Traffic*, *Town Planning Review* 38, no. 2 (July 1967): 156–9.

71 *Swinton Journal* (Summer 1968), 6–7, cited in Cockett, 194 (italics added).

72 Although on the reality of social democracy in Britain see Patrick Joyce, *The State of Freedom: A Social History of the British State since 1800* (Cambridge: Cambridge University Press, 2013), 324–6.

73 Joe Moran, 'Mass-Observation, Market Research, and the Birth of the Focus Group, 1937–1997', *Journal of British Studies* 47, no. 4 (2008): 842; [Peter Hall], *Transport Is Everyone's Problem* (London: *Socialist Commentary*, 1963), i; on the history of the journal see Lawrence Black, 'Social Democracy as a Way of Life: Fellowship and the Socialist Union, 1951–9', *Twentieth Century British History* 10, no. 4 (1999): 499–539.

74 [Hall], *Transport Is Everyone's Problem*, xxxii.

75 [Hall], *Transport Is Everyone's Problem*, xxxviii.

76 Stedman Jones, 9–10.

77 See William Plowden, *The Motor Car and Politics in Britain 1896–1970* (London: Bodley Head, 1971), 346–63.

78 See Thomson.

79 *The Times*, 17 October 1966, 7.

80 Ministry of Transport, *Better Use of Town Roads: The Report of a Study of the Means of Restraint of Traffic on Urban Roads* (London: HMSO, 1967); quotations on 4; on the anti-pricing lobby, see Starkie, 46; and on parking in London see Starkie, Chapter 6.

81 Gunn, 535.

82 Plowden, 401–5, quotation on 401.

83 See Geoff Vigar, *The Politics of Mobility: Transport, the Environment and Public Policy* (London: Spon Press, 2002), 50–2.

84 Starkie, 92–106.
85 For historical accounts of the London congestion charge, see in particular Richards; but also Georgina Santos, 'London Congestion Charging', *Brookings–Wharton Papers on Urban Affairs* (2008), 177–234; Jonathan Leape, 'The London Congestion Charge', *Journal of Economic Perspectives* 20, no. 4 (Fall 2006): 157–76; and Michèle Dix, *The Central London Congestion Charging Scheme: From Conception to Implementation* (London: Transport for London, 2002).
86 Vigar, esp. Chapters 3 and 4; Phil Goodwin, Sharon Hallett, Francesca Kenny, and Gordon Stokes, *Transport: The New Realism* (Oxford: Transport Studies Unit, University of Oxford, 1991); Susan Owens, 'From "Predict and Provide" to "Predict and Prevent"? Pricing and Planning in Transport Policy', *Transport Policy* 2, no. 1 (1995): 43–9.
87 World Commission on Environment and Development, *Our Common Future* (Oxford: Oxford University Press, 1987); and see Vigar, 6.
88 Department of Transport, *National Road Traffic Forecasts* (London: HMSO, 1989); Department of Transport, *Roads for Prosperity* (London: HMSO, 1989); Goodwin *et al.*
89 Goodwin *et al.*, Chapter 7.
90 Patricia Hewitt, *A Cleaner, Faster London: Road Pricing, Transport Policy and the Environment* (London: Institute for Public Policy Research, 1989), 7; Adam Smith Institute transport proposals included Gabriel J. Roth and Eamonn Butler, *Private Road Ahead* (London: Adam Smith Institute, 1982); Eamonn Butler, ed., *Roads and the Private Sector* (London: Adam Smith Institute, 1982); Gabriel J. Roth and Anthony Shephard, *Wheels within Cities: Private Alternatives to Public Transport* (London: Adam Smith Institute, 1984); Madsen Pirie, ed., *Traffic in the City: A Colloquium Sponsored by the Adam Smith Institute* (London: Adam Smith Institute, 1989); and, after Hewitt's IPPR paper, John Hibbs and Gabriel J. Roth, *Tomorrow's Way: Managing Roads in Free Society* (London: Adam Smith Institute, 1992).
91 Cockett, 327–8.
92 Pirie, 24–8; Michael Schabas, *Charging for Roads: A Better Way to Ease Congestion* (London: Centre for Policy Studies, 1995).
93 Richards.
94 Labour Party, *Moving Britain into the 1990s: Labour's New Programme for Transport* (London: Labour Party, 1989); cited in Goodwin *et al.*, 88.
95 MVA Consultancy and Government Office for London, *The London Congestion Charging Research Programme: Principal Findings* (London: HMSO, 1995).
96 United Nations Conference on Environment and Development, *Report of the United Nations Conference on Environment and Development: Rio de Janeiro, 3–14 June 1992* (New York: United Nations, 1993); Royal Commission on Environmental Pollution, *Transport and the Environment (Eighteenth Report)* (London: HMSO, 1994).
97 Department of Transport, *Transport: The Way Forward* (Green Paper) (London: HMSO, 1996); cited in Fiona Poole, *Research Paper 98/16: Traffic Congestion* (London: House of Commons Library, 1998), 36.
98 Poole, 5.
99 Department of the Environment, Transport and the Regions, *A New Deal for Transport: Better for Everyone* (White Paper on integrated transport) (London: The Stationery Office, 1998), Chapter 4.
100 Government Office for London, *Road Charging Options for London: A Technical Assessment* (London: The Stationery Office, 2000).
101 Greater London Authority, *Hearing London's Views: A Discussion Paper on the Mayor's Proposals for Congestion Charging in Central London* (London: Greater London Authority, 2000).
102 Compare for instance Greater London Authority, 1; and Hewitt, 5.
103 Tony Travers, *The Politics of London: Governing an Ungovernable City* (Basingstoke: Palgrave Macmillan, 2004), 190; see also Dave Wetzel, *The London Congestion*

Charge Scheme (New York: Robert Schalkenbach Foundation, 2012), http://schalken bach.org/papers/London-Congestion-Charge-Scheme-RSF-Dave-Wetzel[10].pdf.

Bibliography

The National Archives: Department of Scientific and Industrial Research Papers.

The National Archives: Ministry of Transport Papers.

Hansard.

The Times.

Adshead, Stanley. *A New England: Planning for the Future*. London: Frederick Muller, 1941.

Beckmann, Martin J., Charles Bartlett McGuire, and Christopher Winsten. *Studies in the Economics of Transportation*. New Haven: Yale University Press, 1956.

Beesley, Michael E. Review of D. J. Reynolds, *Economics, Town Planning and Traffic. Economica* 34, no. 136 (November 1967): 451.

Beesley, Michael E., and Gabriel J. Roth. 'Restraint of Traffic in Congested Areas'. *Town Planning Review* 33, no. 3 (1 October 1962): 184–96.

Black, Lawrence. 'Social Democracy as a Way of Life: Fellowship and the Socialist Union, 1951–9'. *Twentieth Century British History* 10, no. 4 (1999): 499–539.

Buchanan, James M. *The Inconsistencies of the National Health Service*. London: Institute of Economic Affairs, 1965.

———. 'The Pricing of Highway Services'. *National Tax Journal* 5, no. 2 (June 1952): 97–106.

Butler, Eamonn, ed. *Roads and the Private Sector*. London: Adam Smith Institute, 1982.

Button, Kenneth. 'The Rationale for Road Pricing: Standard Theory and Latest Advances'. *Research in Transportation Economics* 9 (2004): 3–25.

Christ, Carl F. 'The Cowles Commission's Contributions to Econometrics at Chicago, 1939–1955'. *Journal of Economic Literature* 32, no. 1 (1 March 1994): 30–59.

Coburn, T., Michael Beesley, and Dennis Reynolds. *The London–Birmingham Motorway: Traffic and Economics*. RRL Technical Paper 46. Harmondsworth: Road Research Laboratory, 1960.

Cockett, Richard. *Thinking the Unthinkable: Think-Tanks and the Economic Counter-Revolution, 1931–83*. London: HarperCollins, 1994.

Department of the Environment, Transport and the Regions. *A New Deal for Transport: Better for Everyone* (White Paper on integrated transport). London: The Stationery Office, 1998.

Department of Transport. *National Road Traffic Forecasts*. London: HMSO, 1989.

———. *Roads for Prosperity*. London: HMSO, 1989.

———. *Transport: The Way Forward* (Green Paper). London: HMSO, 1996.

Derycke, Pierre-Henri. 'Road Pricing: Analysis and Policies (a Historical Perspective)'. *Recherches économiques de Louvain* 64, no. 1 (1998): 63–74.

Dix, Michèle. *The Central London Congestion Charging Scheme: From Conception to Implementation*. London: Transport for London, 2002.

Ellis, Howard, and William Fellner. 'External Economies and Diseconomies'. *American Economic Review* 33, no. 3 (September 1943): 492–511.

Evans, Andrew. 'Road Congestion Pricing: When Is It a Good Policy?'. *Journal of Transport Economics and Policy* 26, no. 3 (September 1992): 213–43.

Foster, Christopher. 'Michael Beesley and Cost Benefit Analysis'. *Journal of Transport Economics and Policy* 35, no. 1 (1 January 2001): 3–30.

———. *The Transport Problem*. London: Blackie and Son, 1963.

Friedman, Milton. *A Theory of the Consumption Function*. Princeton: Princeton University Press, 1957.

Gilbert, Dennis. Review of D. J. Reynolds, *Economics, Town Planning and Traffic*. *Town Planning Review* 38, no. 2 (July 1967): 156–9.

Goodwin, Phil. 'Solving Congestion: Inaugural Lecture for the Professorship of Transport Policy, University College London', 23 October 1997. eprints.ucl.ac.uk/1244/1/2004_22.pdf.

Goodwin, Phil, Sharon Hallett, Francesca Kenny, and Gordon Stokes. *Transport: The New Realism*. Oxford: Transport Studies Unit, University of Oxford, 1991.

Government Office for London. *Road Charging Options for London: A Technical Assessment*. London: The Stationery Office, 2000.

Greater London Authority. *Hearing London's Views: A Discussion Paper on the Mayor's Proposals for Congestion Charging in Central London*. London: Greater London Authority, 2000.

Gunn, Simon. 'The Buchanan Report, Environment and the Problem of Traffic in 1960s Britain'. *Twentieth Century British History* 22, no. 4 (2011): 521–42.

Hall, Peter. 'The Buchanan Report: 40 Years On'. *Proceedings of the Institution of Civil Engineers: Transport* 157, no. 1 (February 2004): 7–14.

Hall. *Transport Is Everyone's Problem*. London: Socialist Commentary, 1963.

Hewitt, Patricia. *A Cleaner, Faster London: Road Pricing, Transport Policy and the Environment*. London: Institute for Public Policy Research, 1989.

Hibbs, John, and Gabriel J. Roth. *Tomorrow's Way: Managing Roads in Free Society*. London: Adam Smith Institute, 1992.

Jackson, Ben. 'The Think-Tank Archipelago: Thatcherism and Neo-Liberalism'. In *Making Thatcher's Britain*, ed. Ben Jackson and Robert Saunders, 43–61. Cambridge: Cambridge University Press, 2012.

Joyce, Patrick. *The State of Freedom: A Social History of the British State since 1800*. Cambridge: Cambridge University Press, 2013.

Knight, Frank H. 'Some Fallacies in the Interpretation of Social Cost'. *Quarterly Journal of Economics* 38, no. 4 (August 1924): 582–606.

Labour Party. *Moving Britain into the 1990s: Labour's New Programme for Transport*. London: Labour Party, 1989.

Leape, Jonathan. 'The London Congestion Charge'. *Journal of Economic Perspectives* 20, no. 4 (Fall 2006): 157–76.

Lindsey, Robin. 'Do Economists Reach a Conclusion on Road Pricing? The Intellectual History of an Idea'. *Econ Journal Watch* 3, no. 2 (May 2006): 292–379.

Ministry of Transport. *Better Use of Town Roads: The Report of a Study of the Means of Restraint of Traffic on Urban Roads*. London: HMSO, 1967.

———. *Road Pricing: The Economic and Technical Possibilities*. London: HMSO, 1964.

———. *Traffic in Towns: A Study of the Long Term Problems of Traffic in Urban Areas*. London: HMSO, 1963.

Moran, Joe. 'Mass-Observation, Market Research, and the Birth of the Focus Group, 1937–1997'. *Journal of British Studies* 47, no. 4 (2008): 827–51.

Morrison, Steven. 'A Survey of Road Pricing'. *Transportation Research A* 20, no. 2 (1986): 87–97.

Muller, Christopher. 'The Institute of Economic Affairs: Undermining the Post-War Consensus'. *Contemporary British History* 10, no. 1 (1996): 88–110.

MVA Consultancy, and Government Office for London. *The London Congestion Charging Research Programme: Principal Findings*. London: HMSO, 1995.

Newbery, David, and Georgina Santos. 'Road Taxes, Road User Charges, and Earmarking'. *Fiscal Studies* 20, no. 2 (1999): 103–32.

Overtveldt, Johan van. *The Chicago School: How the University of Chicago Assembled the Thinkers who Revolutionized Economics and Business*. Chicago: Agate, 2007.

Owens, Susan. 'From "Predict and Provide" to "Predict and Prevent"? Pricing and Planning in Transport Policy'. *Transport Policy* 2, no. 1 (1995): 43–9.

Peck, Jamie. *Constructions of Neoliberal Reason*. Oxford: Oxford University Press, 2010.

Pigou, Arthur C. *The Economics of Welfare*. London: Macmillan, 1920.

Pirie, Madsen, ed. *Traffic in the City: A Colloquium Sponsored by the Adam Smith Institute*. London: Adam Smith Institute, 1989.

Plowden, William. *The Motor Car and Politics in Britain 1896–1970*. London: Bodley Head, 1971.

Poole, Fiona. *Research Paper 98/16: Traffic Congestion*. London: House of Commons Library, 1998.

Reynolds, Dennis. 'Congestion'. *Journal of Industrial Economics* 11, no. 2 (April 1963): 132–40.

———. *Economics, Town Planning and Traffic*. London: Institute of Economic Affairs, 1966.

Richards, Martin G. *Congestion Charging in London: The Policy and the Politics*. Basingstoke: Palgrave Macmillan, 2004.

Rooney, David. 'The Political Economy of Congestion: Road Pricing and the Neoliberal Project, 1952–2003'. *Twentieth Century British History* 25, no. 4 (December 2014): 628–50.

Roth, Gabriel J. 'An End to Traffic Jams'. *Crossbow* (September 1962), 13–18.

———. 'A Pricing Policy for Road Space in Town Centres'. *Journal of the Town Planning Institute* (November 1961), 287–9.

———. *A Self-Financing Road System*. London: Institute of Economic Affairs, 1966.

———. *Paying for Parking*. London: Institute of Economic Affairs, 1965.

———. *Roads in a Market Economy*. Aldershot: Avebury Technical, 1996.

Roth, Gabriel J., and Eamonn Butler. *Private Road Ahead*. London: Adam Smith Institute, 1982.

Roth, Gabriel J., and Anthony Shephard. *Wheels within Cities: Private Alternatives to Public Transport*. London: Adam Smith Institute, 1984.

Roth, Gabriel J., and J. Michael Thomson. 'Road Pricing: A Cure for Congestion?'. *Aspect* 3 (1963): 7–14.

Royal Commission on Environmental Pollution. *Transport and the Environment (Eighteenth Report)*. London: HMSO, 1994.

Santos, Georgina. 'London Congestion Charging'. *Brookings–Wharton Papers on Urban Affairs* (2008), 177–234.

Schabas, Michael. *Charging for Roads: A Better Way to Ease Congestion*. London: Centre for Policy Studies, 1995.

Smeed, Reuben. *The Traffic Problem in Towns*. Manchester: Manchester Statistical Society, 1961.

Smith, David. 'Walters, Sir Alan Arthur (1926–2009)'. *Oxford Dictionary of National Biography*, January 2013.

Starkie, David. *The Motorway Age: Road and Traffic Policies in Post-War Britain*. Oxford: Pergamon Press, 1982.

Stedman Jones, Daniel. *Masters of the Universe: Hayek, Friedman, and the Birth of Neoliberal Politics*. Princeton: Princeton University Press, 2012.

Tanner, J. C. *A Theoretical Study of the Possible Use of Tolls to Relieve Traffic Congestion in Urban Areas*. Road Research Laboratory RN/3819/JCT, 1960.

Thomson, J. Michael. 'An Evaluation of Two Proposals for Traffic Restraint in Central London'. *Journal of the Royal Statistical Society Series A (General)* 130, no. 3 (1967): 327–77.

Travers, Tony. *The Politics of London: Governing an Ungovernable City*. Basingstoke: Palgrave Macmillan, 2004.

Tribe, Keith. 'Britain, Economics in (20th Century)'. In *New Palgrave Dictionary of Economics*, 2nd edn. Basingstoke: Palgrave Macmillan, 2008.

———. 'Liberalism and Neoliberalism in Britain, 1930–1980'. In *The Road from Mont Pèlerin: The Making of the Neoliberal Thought Collective*, ed. Philip Mirowski and Dieter Plehwe, 68–97. Cambridge, MA: Harvard University Press, 2009.

United Nations Conference on Environment and Development. *Report of the United Nations Conference on Environment and Development, Rio de Janeiro, 3–14 June 1992*. New York: United Nations, 1993.

Vickrey, William. 'Evidence'. In *Transportation Plan for the National Capital Region: Hearings before the Joint Committee on Washington Metropolitan Problems (86th Congress)*, ed. US Congress, 454–585. Washington: US Congress, 1959.

———. *The Revision of the Rapid Transit Fare Structure of the City of New York*. Technical Monograph 3. New York: Mayor's Committee on Management Survey of the City of New York, 1952.

———. 'Statement to the Joint Committee on Washington, DC, Metropolitan Problems'. *Journal of Urban Economics* 36 (1994): 42–65.

Vigar, Geoff. *The Politics of Mobility: Transport, the Environment and Public Policy*. London: Spon Press, 2002.

Walters, Alan. *The Economics of Road User Charges*. Washington, DC: International Bank for Reconstruction and Development, 1968.

———. 'Empirical Evidence on Optimum Motor Taxes for the United Kingdom'. *Journal of the Royal Statistical Society, Series C (Applied Statistics)* 10, no. 3 (1 November 1961): 157–69.

———. *Integration in Freight Transport*. London: Institute of Economic Affairs, 1968.

———. 'A Life Philosophy'. *American Economist* 33, no. 2 (1 October 1989): 18–24.

———. 'The Theory and Measurement of Private and Social Cost of Highway Congestion'. *Econometrica* 29, no. 4 (1 October 1961): 676–99.

———. *The Theory and Measurement of Private and Social Cost of Highway Congestion (Discussion Paper)*. Faculty of Commerce and Social Science Discussion Papers B1. Birmingham: University of Birmingham, 1959.

———. 'Track Costs and Motor Taxation'. *Journal of Industrial Economics* 2, no. 2 (1 April 1954): 135–46.

Wetzel, Dave. *The London Congestion Charge Scheme*. New York: Robert Schalkenbach Foundation, 2012. http://schalkenbach.org/papers/London-Congestion-Charge-Scheme-RSF-Dave-Wetzel[10].pdf.

Winsten, Christopher. Review of D. J. Reynolds, *Economics, Town Planning and Traffic*. *Journal of Transport Economics and Policy* 1, no. 1 (January 1967): 118–19.

World Commission on Environment and Development. *Our Common Future*. Oxford: Oxford University Press, 1987.

6 Scientists, sensors, and surveillance

Introduction

This chapter opens out the discourse of electronic networks touched on in the previous chapter, and widens the network of practitioners interested in the opportunities that electronic sensors and computers offered in measuring, understanding, and controlling the traffic network (in the socio-technical sense that runs throughout this book). It explores a series of 1960s experiments in London and Glasgow on computer-controlled networks of vehicle sensors, closed-circuit television (CCTV) cameras, and traffic lights running pattern-recognition software. This was the work of a range of practitioners including statisticians, modellers, computer programmers, systems analysts, and electronic engineers, collectively termed 'scientists' to recognize an approach based on experimentation and the manipulation of data. This approach combined measurement (gathered by sensors and surveillance) with models (run on electronic computers) to simulate traffic problems as well as to respond to real situations on the ground. It thus embodied network ideas of traffic into and over the streets as well as in computer networks; traffic, in this characterization, was conceived more from the atomic level of the single vehicle upwards, than from the holistic planner's top-down conception of aggregate flows.

In 1971, the *Times* technology editor, Kenneth Owen, reported on the results of the experiments, describing a 'computerized way out of the clogged streets' in which advanced electronics, combined with developments in cybernetics, could offer new methods to control road traffic across cities. Automation on the streets of London was fast approaching, through a combination of technology and mathematics – all with the express intention of unclogging the traffic.[1] As journalist Tony Aldous later put it, the traffic problem in London was being solved by 'six computers and a room full of cathode ray tubes and policemen just off Victoria Street'.[2]

What did the computer offer to London's traffic problem to enable it to become so central, so quickly, to the story of congestion? The answer came in the form of systems analysis – a mathematical approach to understanding and controlling a complex interacting network in its entirety. But it was also political. We have seen, time and again, a desire, in those who have sought to solve congestion, for control of the traffic over a wide area of London – a desire to look beyond the local and see London as a whole system that could be analysed, understood, and controlled. Local schemes were all very well but the effect on the whole was never clearly understood, nor clearly enough under the control of a single body. The digital computer, connected to a network of sensors and cameras, appeared to offer both those facilities, and went on to grow profoundly over subsequent decades. This chapter will chart its growth and its relationship with cultures of control and surveillance in the capital. The first trial of the new technology was a prime London location, comprising the districts of Knightsbridge, Chelsea, South Kensington, and Hammersmith, in a project known as the West London Traffic Experiment.

The West London Traffic Experiment, 1963–1968

By the 1960s, it was generally held that London's traffic had become dire.[3] One Ministry of Transport official summarized the problems in a speech as follows:

> The remorseless growth of road traffic, the resulting overloading of urban road networks, the congestion, the frustrations and delays that result: all these effects are there in our cities for everyone to see. The cost of the delays to the community is enormous. The cost of new roads etc., and of comprehensive redevelopment to provide for extra traffic is so high, and imposes such strains on the civil engineering and building industries that in practice progress cannot keep pace with traffic growth. Then there is also the danger that wholesale rebuilding may destroy the essential character of our cities.

What was the solution? The Ministry engineer continued as follows:

> In these circumstances it is essential to obtain the utmost efficiency from existing streets. Much can be done, of course, by traffic engineering measures – by one-way streets, by banned turns, by waiting restrictions and so on but when all is done that reasonably can be done by such means we are still left with congestion and delay. One major source of delay which remains is delay at signal controlled junctions.[4]

The idea of getting more capacity from the existing road network – traffic management – had been given practical form in 1960 with the establishment by new transport minister Ernest Marples of the London Traffic Management Unit, following a study visit by him to the USA. Within months, a swathe of new projects was under way in the capital, including the Tottenham Court Road and Gower Street one-way network, the creation of box junctions, an increase in lane markings and traffic signs, and countless minor physical alterations to road widths and corners.[5] Traffic management was a synthesis of earlier approaches that we have examined already, including traffic control, engineering, and planning. But, as the Ministry of Transport official quoted above explained, one problem remaining to be solved was the flow of vehicles through junctions, and addressing delays at intersections in dense urban street networks demanded improvements in two areas: firstly, a better understanding of the way traffic flowed, and secondly, a better system of timing traffic signals to respond to changing traffic conditions.

Proposals for computer control of traffic signalling over a wide urban area – known as area traffic control – dated back as early as 1950, with the Institute of Transportation and Traffic Engineering in Los Angeles proposing a computerized traffic sensor network operating tidal flow lanes and signs showing optimum speeds. In 1954, the same group proposed an elaborate computer simulation of an area-wide signal network.[6] Other ideas followed. Yet in the UK it was the agitation of electronics and defence firms in 1960, keen to generate new business in Harold Macmillan's lucrative spending market, that prompted the Ministry of Transport to examine the matter in detail, asking firms to submit proposals.[7] Ernest Marples took particular interest in area traffic control and, in March 1961, promised to support research – one idea being for the government's Road Research Laboratory (RRL) to 'be given the Western Approaches to London on which to experiment'.[8]

It got its way. A panel was set up to investigate innovations in traffic control, as well as to propose a forward programme of research.[9] Its July 1962 report recommended a trial installation of new technologies and, in 1963, a 6.5-square-mile area of west London, containing 100 sets of traffic lights, was selected for a £550,000 Ministry of Transport-led experiment in area traffic control.[10] The aim was to bring traffic signals under computer control in three ways: firstly, by altering the timing of each signal according to traffic flow models responding to changing traffic patterns; secondly, by diverting traffic away from congested areas; and thirdly, to operate tidal flow lanes for morning and evening peaks.[11]

The west London area was chosen, it was said, as a 'commuter corridor' that included 'some of the busiest roads in London servicing museums, concert halls, exhibitions, a football ground and major shopping centres and hotels as well as the main route to and from London Airport', and the experiment involved 'a configuration of data processing and traffic control equipment unique anywhere in the world'.[12] Figure 6.1 shows the area covered.[13] Yet, in reality, it was chosen *not* because its roads were especially congested. The area lent itself to the trial because it was fairly close to the RRL site near Heathrow Airport; it had enough signals to allow extensive trials; it had limited numbers of roads in and out of the area; it had several main thoroughfares, enabling quick assessment of equipment; it had wide variation in traffic patterns; and, as the RRL put it, 'traffic is relatively free in the area', although it observed that intense congestion 'will probably have to be tackled at a later stage'.[14]

The ultimate objective of area traffic control was to reduce the total journey time of all traffic passing through the area to a minimum. This would increase capacity on the road network, especially on through-routes, and was partly designed in order to reduce the incidence of motorists using residential side-

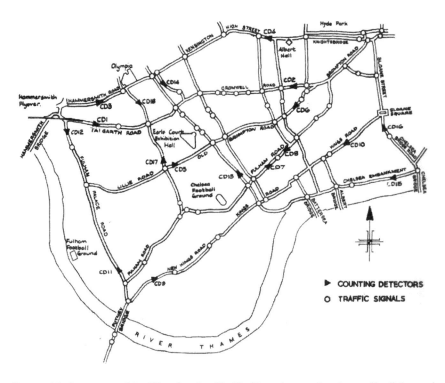

Figure 6.1 Area covered by West London Traffic Experiment, showing traffic lights and queue detectors (Ministry of Transport, Crown Copyright, reproduced under Open Government Licence v3.0).

roads as rat-runs to avoid congestion. Co-ordination of traffic lights along a particular street was not new, as was shown in Chapter 4. Locally interlinked traffic light networks had been installed on London's roads as far back as the 1930s. But the rise, during and after the Second World War, of the electronic computer and developments in fields such as operations research and systems analysis offered a new opportunity, namely to understand and shape the behaviour of traffic across a whole area.

One Ministry of Transport engineer described the problem of locally interlinked traffic signals, each running to its own timing plan, as follows:

> In an area one route intersects with another, each requiring separate plans at different times of day to minimise the delay, and these routes again are crossed by further routes and so on, each intersecting with the other and forming a series of loops which make the problem of devising one total plan to minimise journey times extremely complex. It is in a situation such as this that the power and flexibility of a computer can be used to the full.[15]

The digital computer enabled a systems approach to traffic congestion, following earlier experiments with remote traffic control at isolated locations including Vauxhall Cross and the elevated section of the M4.[16] Unlike the existing installations, the west London trial treated a wide area of London as a single system in which traffic flows in one area affected those elsewhere, and made use of a network of vehicle detectors connected to the central computer. These sensors measured the volume of traffic in the system, the number of vehicles approaching each junction, traffic speed, the length of queues building up, and the presence of slow-moving or stationary traffic.

Work began in 1964 under the leadership of Ministry of Transport traffic engineer Brian Cobbe, who was given a new department to run in collaboration with the Greater London Council (GLC), RRL, and Metropolitan Police. After tendering, lead contractors Plessey Automation and GEC were appointed to develop the hardware, but early Ministry plans to recruit in-house software- and systems-specialists from within the Civil Service foundered after a fruitless two-year search, which included advertisements in the national press. Instead, the entire project relied on a specialist programming team supplied by the Plessey company itself.[17]

After almost four years of field research, design, planning, and construction, the West London Traffic Experiment went live on 22 January 1968. Inside its New Scotland Yard control room, staffed by police officers, Ministry of Transport operators (trained by Plessey specialists), and a computer manager, two Plessey XL9 computers sat ready to tune the traffic signals: one to handle the second-by-second stream of incoming data and outgoing commands flowing through telephone lines to each intersection, and the other to use pattern recognition to select the best from a library of plans for different traffic situations.[18] When Minister of State Stephen Swingler switched the computer on, 150 miles of London's streets came under computer control.

Cybernetics on the streets of London

The central computer was seemingly the most novel part of the experiment, and the traffic signals at each junction the most visible, but it was the vehicle sensor network that was the most physically complex and costly part of the project, comprising the interconnection of 500 vehicle-sensing loops in the roadway – some new, some from existing vehicle-activated traffic lights – with electronic control equipment in boxes under the pavement or on posts nearby.[19] Altogether, 20 miles of new cabling connected the sensors, control boxes, data out-stations, and pedestrian controllers to the network.[20]

Once installed, the experimental sensor system could detect the number of vehicles flowing through the west London road network in real time. This was not new; existing vehicle-actuated traffic lights had operated using pneumatic sensors for many years. But they could only detect freely moving vehicles. The new system went an important step further. Its sensors could detect the presence of stopped or sluggish traffic. This was a significant moment in the history of

London traffic. For the first time, congestion itself could be sensed automatically, remotely, over a large area of the capital – and, moreover, it could cause its own reduction, in a cybernetic system of feedback and control.[21]

In reality, the west London experiment was (of course) a system with both human and non-human actors, and the central control room in New Scotland Yard was the location for human surveillance of the network. At its heart was a dynamic map display that normally showed an overview of the system – for instance showing the signals currently at green – but on which operators could call up the indication of any traffic light or vehicle sensor in the system at any time.[22] Critical intersections could be called up on mimic panels, including a display of all the light phases available at that junction. Intersections could be switched to manual control directly from the control room, or placed under local controls operated by police officers at the scene. Alarms sounded when abnormal situations developed – such as blockages owing to accidents – or if the computer did not recognize the pattern of traffic behaviour, and in such cases the computer made its best guess but could be overridden.[23]

A critical element of the west London system was the inclusion of eight CCTV cameras mounted high above the streets at key intersections, enabling officers in the control room to observe the traffic live on the streets below. A newsreel film shot as the system was being completed explained:

> The 'traffic eyes' will help to spot and sort out some of the tangles that often come close to completely choking the streets … the high-mounted cameras can take in a large area, and the man at control sees at a glance exactly what's happening, and quickly gets to grips with the overall problem.[24]

We will return to the CCTV aspect of the project presently.

The west London experiment followed developments in several North American cities, which, in the words of a 1968 *Times* editorial entitled 'The Computer is Watching', 'tried to use the computer to break the bottlenecks as motorists pour on to the roads at peak hours'.[25] But the UK incarnation was more complex, being described by one Ministry of Transport engineer as 'probably the most advanced application of process control by computers which has yet been attempted', and by another as 'almost frightening in its complexity', involving not just engineering but 'planning and social issues'.[26]

Minister Swingler was at pains to point out the experimental nature of the west London scheme and the sophisticated system it was designed to assist. Observing that London had long had traffic lights that responded to vehicle levels, rather than the American system of fixed-time-cycle lights, he explained that the addition of computer control offered 'less room for improvement, less slack to be taken up', and that 'the problem of achieving greater efficiency is that much harder'. He was managing the expectations of motorists and journalists, noting that:

> there are going to be no immediate, dramatic results … we shall be learning as we go along … we shall be checking the delays and journey times of

vehicles travelling through the area and this information will be going into the computers for analysis.[27]

Even preliminary findings would take months.

In part, the politicians were downplaying the potential of the scheme because they were still wedded to the reconstruction schemes explored in Chapter 2; success with computerized traffic management would diminish the perceived need for new roads. But this was a genuinely new technological field of a hitherto unknown complexity. The key issue raised by the 1968 *Times* editorial concerned what the paper called the 'social problem between the motorist and the traffic planner', namely whether the human actors in the network would follow the rules or not. Four years previously, following area traffic control experiments in Toronto, the newspaper had described a future in which the computer was a 'road "overlord" ... squatting in a central office' where it would 'rule the cars by altering traffic light sequences, and directing traffic down side streets when jams begin to build up'.[28] The west London scheme's head, Brian Cobbe, stated that 'as long as motorists obey the traffic lights, the computer will do the rest. The system's success depends on obedience, laced with common sense'.[29]

But for Swingler, there was no issue: the cameras and sensors were merely extensions of the human system already established to control London's streets. 'Computers are not robots and they need the human touch', he remarked. 'Essentially it is the police and the traffic engineers in the control room and the police on the ground in West London who will provide this.'[30] A spokesman for Elliott Automation, a traffic control equipment manufacturer, had explained that computers could 'never perform calculations beyond the capacity of a small child', but could perform them very quickly, concluding that 'the computer is a moron, but a very high speed moron at that'.[31]

But the cybernetic computerized system was far more than merely a high-speed calculator. It represented a fundamental shift towards a world view in which vehicles could be reduced to computer bits, and traffic reduced to probabilities – with a concomitant shift in the *location* of congestion. It was leaving the streets and moving to the humming computers at New Scotland Yard. We shall now turn to the modelling and simulation of congestion.

Modelling and simulating congestion

The technologies and practices required to enable large-scale modelling and simulation of traffic in networks such as London were not available until the 1950s, with programmable computers, operations research, and systems analysis. Linked to this were ideas about mathematically modelling cities in general that emerged from the schools of urban planning and geography. Together, this combination of high-speed digital computation, systems analysis, post-war urban planning, and the quantitative turn towards spatiality in geography nurtured a culture of positivist modelling in which cities could be reduced to idealized models, vehicles to atomistic computer digits, and the decisions of road-users to statistical probabilities. The 1960s thus saw a huge increase in research into the science and mathematics of traffic, one 1964 bibliography comprising over 1,000 papers relating to the topic.[32]

A typical example came in 1961, when the traffic statistician Reuben Smeed, Deputy Director of the RRL, who was by then getting involved with road pricing as discussed in Chapter 5, published a mathematical model of a theoretical town in a bid to understand the space requirements of its ideal road system. He introduced it as follows:

> It will be supposed that the town is circular and symmetrical about the centre. All the N workers work within a radius r of the town centre and live between circles of radii r and R. Except in the space required for roads, the work places are uniformly distributed in the central part of the town, within the radius r and their houses are uniformly distributed over the rest of the town between the circles radii r and R. In his journey to work, the worker travels along the straight line from his home to the town centre until he reaches radius r. He then travels directly towards the place at which he works, the position of which is not correlated with the place at which he lives or with the point at which he reaches the outer part of the town centre.[33]

After 36 pages of mathematical calculations, one of Smeed's conclusions was that 'in cases where the central area is concentrated, and the numbers of persons large, it would not be possible to enable every person to travel to work by car unless roads were built on top of one another'.[34]

Of course, Smeed was aware that the assumptions in his model did not apply to a real town, but he concluded that it was useful enough to meet his aim of calculating road space requirements of different forms of transport. 'Congestion', he exclaimed, 'is rife during large portions of the day in many urban areas. Road traffic is reduced to a crawl, passengers crowd into trains until no more can be packed in. The urban areas themselves are becoming larger and larger.'[35] He simply wanted to offer a comprehensive quantitative analysis of the problem, and continued to work on the problem throughout the 1960s.[36]

Much has been written about computer urban modelling in this period, often focusing on combined land-use–transport models such as Ira Lowry's 1964 model of Pittsburgh, USA, which sought to understand the spatial generators and attractors of vehicle journeys in cities as part of the urban planning process.[37] In fact, it was early successes in computerized traffic models in the 1950s that led to large-scale modelling of the urban environment, as Lowry himself acknowledged in his study.[38] As Michael Batty put it, 'the idea of urban planning as "architecture-writ-large" waned, as systems, models and computers became the fashion' in the 1950s and 1960s.[39] Theirs was not the only world view, however; the reductionist approach of the modellers attracted early and trenchant criticism from social theorists such as David Harvey and Andrew Sayer who, in different ways, questioned the entire basis of abstracted modelling and its assumptions of efficiency.[40] Yet, however worthwhile their practice, traffic simulation and modelling represented a paradigm shift of thinking in this period.[41]

An early example of simulating traffic using a digital computer was Daniel Gerlough's 1955 doctoral book at the University of California.[42] The popular science periodical *Science News Letter* was captivated by the possibilities of Gerlough's model, exclaiming 'Electronic "Brain" Can Solve Traffic Problems' before going on to describe 'a miniature highway intersection simulated inside the metallic guts of an electronic computer' that was to study traffic flow at highway junctions. By today's standards it was rudimentary, as the following description makes clear:

> Electrical impulses, each representing a car, are 'stopped', make right and left turns, obey 'traffic signals' and watch out for 'pedestrians' and 'other cars' in the maze of circuits ... Such electronic 'brains' could help designers find the most efficient pattern for a specific intersection.[43]

Gerlough's work in the USA was just predated by research carried out at the UK's RRL, first on an electro-mechanical computer and then on Alan Turing's Pilot ACE computer at the National Physical Laboratory, which simulated a four-way junction at six times life speed.[44] Both simulations were limited, however, by a focus on a single intersection rather than a network of roads.

Nevertheless, this computer simulation work was taking place alongside research into the statistical mathematics of congestion and queuing, which meant that the methodology was ready as soon as the hardware speed caught up (which happened rapidly in the 1950s).[45] By the early 1960s, as the west London experiment was being established, work on computer simulation of complex multi-lane urban road networks was being routinely carried out, such as a 19-road, four-lane network comprising nine intersections (based on a real layout in Manchester) that was modelled on a Mercury computer at Manchester University in 1962.[46] The following year, Daniel Gerlough published details of a simulation of 80 intersections and 217 road links.[47]

What was at stake here was not simply a better understanding of traffic in cities. Through mathematical models and computer simulation, the systems analysts were creating virtual cities, withdrawing into a world in which experimentation could take place far from the dirt and politics of the real streets. They were also creating a scientific discourse of traffic. An IBM analyst explained the role of simulation as a means of representation as follows:

> Airline pilots receive training and experience through simulated flights without actually flying the real aeroplanes ... simulation is like experimentation which is a primary tool of chemistry, physics and other sciences. In the study of vehicular traffic, however, experimentation is not always possible. There are too many variables to allow experimental study under a rigorously controlled environment without enormous expenditure and

unbearable inconvenience to the motoring public ... The experience and knowledge gained from simulation studies can be used to construct more sophisticated mathematical models and control strategies, which in turn provide a better simulation.[48]

The west London scheme did not incorporate the latest in modelling and simulation. It was a scheme in which the computer responded to real traffic by selecting from a library of timing plans written by traffic engineers. It was strongly driven by political concerns and a requirement to unclog the capital's real streets as quickly as possible as part of Marples's London traffic management project. Yet, given this backdrop of over a decade of academic research into mathematical modelling and computer simulation, it is not surprising that the mathematicians, programmers, and systems analysts wanted to go further in distancing the computer from the human world – they wanted to experiment, but politics dictated that London was not the place to do so. But far away from Marples and the controlling hand of the Ministry of Transport, the RRL had a second project.

Glasgow, the digital city

In Glasgow, the RRL began a series of experiments in November 1967 in which computer simulations of traffic flowing around the central area under five different mathematical models were compared with reality in an elaborate computerized network of lights, sensors, and controllers.[49] If west London was primarily about rapid calculation, using a computer in a quick empirical trial, Glasgow was about experimentation: abstract congestion in an idealized city.[50] In the Scottish installation, 80 traffic signals covering 1 square mile of the city were brought under computer control, compared with west London's 100 signals in 6.5 square miles.[51]

Why was Glasgow chosen for this elaborate series of experiments? One reason, as has already been suggested, was that it was far away from Westminster politics. The RRL wanted a test-bed and time to carry out extensive tests, and internal departmental correspondence shows that the transport ministry cared little for any experiment outside London; it was happy to let Glasgow go ahead so long as progress on west London was not jeopardized.[52] But another reason was that Glasgow had particular characteristics that suited the RRL's requirements. Its City Corporation had expressed enthusiasm in experimenting with area traffic control. The RRL's Scottish Branch was nearby. And the city centre was compact with a high density of traffic lights. These were the reasons given publicly for its selection.[53] But there were two further characteristics of Glasgow's road network that are significant in this study of London's traffic problem.

The first was its layout, which, as Figure 6.2 shows, takes a grid form.[54] We have seen in Chapter 3 how engineer Charles Bressey wanted to overlay a concrete grid over the existing streets, and in Chapter 4 how Traffic Commissioner Alker Tripp sought to impose grid-like behaviour onto London's unruly street pattern using segregation. For the RRL, 1960s Glasgow seemed to offer a

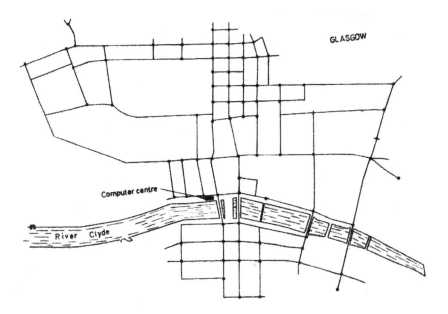

Figure 6.2 Network of traffic signals in Glasgow controlled by computer, 1968 (Ministry of Transport, Crown Copyright, reproduced under Open Government Licence v3.0).

real grid network on which to experiment – a layout that could readily be modelled using the computer technology of the day.

Yet, as with London, Glasgow refused to conform to classical grid behaviour despite its similarities with city centres such as Chicago and New York. Road Research Laboratory statistician Joyce Holroyd noted in 1971:

> It has sometimes been suggested that the results obtained in Glasgow owe something to the particularly 'grid-iron' nature of the street layout ... This is not believed to have been a significant factor because steep hills in some parts of the city effectively prevent major through movements.[55]

Grid historian Hannah Higgins has asserted that the street patterns of New York and Chicago were 'not responsive to unique functions and features of the landscape'.[56] That might be true, but, in Glasgow at least, the landscape certainly made its presence felt as vehicles laboured up Douglas Street or took detours around North Portland Street to avoid the treacherous gradients.

In part, Holroyd's comments were an attempt to play down Glasgow's uniqueness, as the RRL intended to export its models, systems, and expertise to cities around the world. But they must also remind us that, where traffic is concerned, cities are experienced as much in elevation as in plan; and, more generally, they must be seen as an admission that specific local conditions will always shape the adoption and operation of global or idealized models. Ideas travel, but are modified.

The second characteristic of Glasgow's road network that shines light onto its selection as an RRL experimental site is the wider context of planning in the city. Here we have further evidence of tension between the Abercrombie-era planning approach discussed in Chapter 2, with its brutal segregated concrete urban rings and arteries, and a new systems approach based on efficiency, models, and flow – in effect, seeing the city as an analogue for a piece of computer software. For Glasgow in the 1960s was one of the sites where the brutalist approach was most advanced and, for some, becoming most contested.

Simon Gunn has observed that, in the 1950s, 'in cities like Birmingham and Glasgow urban motorways were seen as markers of civic ambition', but that the period was 'marked by a deep duality that was to colour both political and popular responses to automobility throughout the 1960s'.[57] Proposals for urban motorways across Glasgow originated in principle in the 1945 Bruce Report, and the Abercrombie Report the following year.[58] From 1960 onwards, these ideas matured into firm plans for a 48-mile city-wide network of motorways, published and approved in 1963, with construction of the first phase – today's M8, which encircles the city centre – starting in November 1965 and being completed 2.5 years later.[59] By 1967, therefore, as the RRL statisticians were setting up their computer control room on Glasgow's quayside, a few blocks away the city was in the process of being torn apart by diggers and road-building machinery, and concerns over the pro-urban-motorway paradigm were beginning to accelerate rapidly as environmental politics was gaining a voice.[60] One of the experiment's engineers remarked that 'At present the rate of increase in road traffic is far greater than the rate at which roads are built to handle it.'[61] Government ministers and transport specialists were sensing a shift in the mood of the populace away from the brutal approach of Abercrombie, as the diggers advanced, and, perhaps unwittingly, the RRL picked a city that was to become a symbolic location of the tension between roads and environment.

This tension becomes all the more marked when one learns that digital computers had been in use since the early 1960s to model traffic along Glasgow's streets – but, unlike the RRL programmers trying to squeeze more capacity from the existing streets, these early programmes were written by consultants working on the city's motorway programme to model predicted flows on the new network.[62] The new electronic networks offered digital futures, but those futures remained ambivalent: they could support or dismantle Abercrombie's dream. Whichever side of the road-building debate they sat on, the programmers and statisticians agreed on one conclusion: that 'we are entering a period where the digital computer will become a tool of major importance in relieving congestion and delays in the cities of today and tomorrow'.[63] Furthermore – in a plotline anticipating the film *WarGames* by some 15 years – they envisaged a time when the computer could learn from its surroundings and adapt its behaviour according to experience.

The politicians agreed – early indications from west London and Glasgow were that the computerized signals were indeed reducing congestion and increasing journey speed. All the signs pointed towards a more ambitious experiment, bringing together simulation, machine-learning, and the politics of London's government. As early as 1966, even before the west London and Glasgow experiments went live, plans began to be put in place to extend area traffic control across London using new computers in an ambitious £3.75 million programme.[64] Together, the west London and Glasgow experiments were to feed directly into a scheme called 'CITRAC' (Central Integrated Traffic Control), London's first city-wide traffic control system, which began to be installed in 1968. This time, though, the nascent GLC was in the driving seat.

CITRAC and the remodelling of London, 1968–1978

In the decade following the launch of the West London Traffic Experiment in 1968, area traffic control grew like Topsy across the capital, in the project that became known as CITRAC. It began formally in December 1968, when approval was given by the GLC to bring 300 traffic signals in London's West End district under computer control. This phase – as well as 40 signals in the City of London – went live in 1972, and was housed in the existing control room at New Scotland Yard along with new computer graphics facilities to supplement the wall map. The following year, a new CITRAC centre was established in the building, incorporating a new police operations room staffed by police officers (replacing the old west London facility, which moved to the new suite); a computer suite staffed by GLC staff; a fault control room also staffed by GLC engineers; and a suite of offices for supervisors, systems analysts, and programmers.

In 1974, a further 150 sets of traffic lights were incorporated into CITRAC covering the area encircling the West End. Two years later, work towards incorporating 500 more lights into the scheme was under way, and the ultimate system of 1,000 traffic lights over 246 square miles of London was completed by about 1978. Three Siemens 306 digital computers, plus a Marconi Myriad II, had joined the original two Plessey XL9s at New Scotland Yard.[65]

CITRAC combined elements of both the west London experiment, which tested technologies of vehicle sensing and rapid computerized calculation, and the Glasgow city-wide scheme, in which mathematical traffic models were simulated in the computer before being tested on the real-life streets.[66] In CITRAC, Glasgow-style computer models were dreamt up by GLC analysts before being tried on London's traffic, meaning that congestion-management strategies could change in real time with shifting traffic policies.

This was a significant development. CITRAC had been born in the 1960s world of the GLC's Ringways. This was a world of capacity expansion, so CITRAC's original models were designed to increase traffic capacity on the new and existing roads – to speed the flow. These included an RRL model tested in Glasgow known as 'TRANSYT', which attempted to create free-flowing traffic whatever the class of vehicle.[67] But, by 1974, as we have seen, road-building in London was all but off the political map, replaced by a new politics of public transport and traffic restraint. In this new policy world, another RRL model known as 'Bus TRANSYT', which gave more green-signal time to public transport at the expense of the private motorist, made more political sense, so the system was switched over to the revised model.[68]

But CITRAC could do more than simply usher a few more buses through each green phase. As well as attempting to smooth traffic flow, it could literally move congestion from one place to another. Its sensors could detect queues forming on bus routes and switch the signal timings in order to shift the queues onto what the police euphemistically termed 'storage space' – in reality, commuter routes such as the Westway that carried no buses. Furthermore, policies to limit the number of private vehicles entering central London were enacted live on the streets by creating real-time 'restraint cordons', signalling incoming traffic onto orbital routes when traffic levels rose.

A police report noted the power of the electronic system over the traditional approach of physical intervention as follows:

> The accommodation of queues and blocking of 'rat-runs' is obviously a difficult environmental and planning matter. The advantage of introducing the restraint by signals would be that it could be controlled to suit traffic conditions and could be made automatically regulating by means of detectors.[69]

By the reprogramming of its models, CITRAC could enact new government policies overnight. If transport policy changed back from restraint to expansion, CITRAC could make the adjustments simply by loading in the old TRANSYT tape during the nightly update.

The systems analysts could therefore do what no urban planners – from Christopher Wren onwards – had ever succeeded in doing. They could reshape London's streets, automatically, second by second. The computer overlord, with its eyes in the sky and its electronic sensors under the streets, was solving the capital's chronic congestion problem without flattening cherished buildings. But it was not just traffic policy that CITRAC could help implement. It was easy to believe that the system was truly automatic – a mechanism for liberal democracies to exert benign control. But what was the system really keeping an eye on? Who was being watched, and why? In particular, in an automatic system of sensors and simulations, why was the control room staffed by Metropolitan Police officers, while the systems owners, the GLC, were relegated to a back room? If congestion was a pathological form of disordered streets, we will now conclude by examining the place of CITRAC in public order.

Surveillance and the creation of ordered streets

We have seen that the West London Traffic Experiment, which went live in January 1968, included eight CCTV cameras offering visual surveillance of critical traffic junctions. But that was nothing compared with CITRAC, which followed it. By the time of its completion in the late 1970s, the city-wide scheme had augmented that modest number of cameras by a further 150 – all monitored in the police control room with facilities to pan, tilt, and zoom the cameras. For the first time, the whole of Greater London's main road network was under the remote gaze of the traffic police, who had recognized that central control would enhance their presence on the streets.

A police briefing on the CITRAC system under construction in 1974 explained the situation as follows:

> In addition to all the sophisticated equipment provided by the project, Area Traffic Control Room staff have direct tactical control of police resources in the street. This ensures that all incidents can be dealt with speedily and efficiently – either by operations room staff alone in some instances, or combined with traffic patrol units in the street when necessary.[70]

But it was not just the traffic cops who could see through the eyes of CITRAC. By examining the rhetoric of congestion, we can see a significant relationship between traffic and the rise of public-area CCTV surveillance in Britain's towns and cities. To do so, we need to track two parallel debates, namely traffic and public order, and the ways they have intersected with each other.

The management committee of the west London trial commented in September 1968, after nine months of the scheme's operation, that there had been 'a few "Big Brother" comments about the closed-circuit television'.[71] In fact, British police forces had begun experimenting with CCTV cameras for traffic congestion reduction as early as 1956, with a small experiment in the northern cathedral city of Durham. After that, a series of CCTV experiments for public-order purposes was carried out by the Metropolitan Police in London starting in 1960, as we shall see presently. Following these early experiments – and at the same time that CITRAC was being installed by the GLC – the Metropolitan Police went on to install its own dedicated CCTV network across London to provide surveillance of political demonstrations and other street activity. But, crucially, these public-order cameras were presented in public as part of the CITRAC traffic scheme.

Most writers have dated the introduction of public-area CCTV in the UK to the 1980s, in a context of Tory policies to address fear of crime.[72] However, Chris Williams has correctly shifted that date some 15 years earlier, examining the installation of the technology by UK police forces in the 1960s in the context of centralized information-gathering and concerns over public disorder, calling partly on the work of Sarah Manwaring-White, who reported on Metropolitan Police activity in the 1960s (including CITRAC) as early as 1983.[73] Writers for the British Society for Social Responsibility in Science described police use of the CITRAC system in 1985, but without dating it to the 1960s.[74] We have already seen that electronics manufacturers were lobbying government departments in 1960 for area traffic control experiments, and Williams notes that at the same time they were also promoting CCTV installations to the Metropolitan

Police, which held an internal exhibition of the technologies. The overt context was traffic control, but this was a Trojan horse, as the original police files betray.

One police superintendent who examined the exhibits made the following observations:

> Traffic problems near the House of Commons, Trafalgar Square and Hyde Park are well nigh a permanent feature today. To relieve the congestion with its resultant saving in man-power, I consider that a constructive step to improvement would be the introduction of TV Closed Circuit system in these areas ... The cameras could be suitably disguised, where necessary ... *The value of this system where political meetings break out in disorder would prove of immense value and enable reserves to be despatched to the exact seat of the trouble with a minimum of delay.*[75]

Another superintendent who attended the exhibition made similar comments:

> I was very impressed with the closed circuit television demonstration ... It was portable, reasonably small, and easily concealed, and *for difficult observation purposes, particularly in the CID field, has I feel considerable possibilities.* It could also have a useful application in the control of difficult traffic junctions and where extensive road works are being undertaken.[76]

Following the exhibition, two trials were held in Trafalgar Square that year using equipment lent by the manufacturer EMI. The first monitored the July state visit of the Thai royal family (widely reported in the national press as a traffic control experiment), with the second observing crowds on Guy Fawkes night.[77] The royal visit was described by the EMI press office as 'believed to be the first time that closed-circuit equipment has been used for such a purpose in this country'.[78] The technical quality of the pictures was disappointing, however, and it took five years until another trial took place, this time to combat theft at Hatton Garden and Ruislip.[79]

A further police experiment in 1965 saw cameras monitoring traffic at two busy junctions in London's Clerkenwell as part of Birmingham University research into traffic control technologies, and a similar installation was erected on Hammersmith Bridge to operate a tidal flow system, part of congestion reduction measures on the nearby Hammersmith gyratory system. There were already CCTV cameras assisting police on the M6 motorway, and the new M4 London section was being similarly fitted out. At this time, though, cameras for crime prevention were not gaining support, partly owing to technical limitations in low light revealed by the Trafalgar Square and Hatton Garden trials.[80]

However, a breakthrough in the public-order debate came in 1968. The previous year, the Metropolitan Police had set up a temporary four-camera CCTV installation in Croydon to help detect crime. Originally, the cameras were to be relocated to Chelsea for further trial, but this plan was abandoned in summer 1968 following widespread disorder at an anti-Vietnam march in

London in a year of escalating political protest and tension worldwide. With a further demonstration due in Grosvenor Square that October, police made plans to reinstall the Croydon cameras to cover the route of the march.[81]

This 1968 experiment was the first to be widely seen by the police as a public-order success, and three of the cameras were made permanent, joined by the erection of two more the following year to cover demonstrations by groups as diverse as Vietnam protestors, Irish republicans, Ulster solidarity campaigners, communists, anti-apartheid protestors, and the Universal Coloured Peoples and Arabs Association.[82] Williams has described this 1968/9 five-camera police CCTV network, covering the heart of government in Whitehall and Parliament, as 'the first permanent surveillance system' in London.[83] However, he has not noted the eight permanent cameras of the west London experiment, which were operational in the Scotland Yard control room by January 1968, nor the ambitious CITRAC plans in place for so many more cameras to be installed through the 1970s. Was there no connection between the CITRAC cameras and the public-order installations?

By 1970, following the apparent success of remote surveillance in policing the political demonstrations of the late 1960s, CCTV was seen by the police as a highly desirable public-order and surveillance tool. The force's commissioner, Sir John Waldron, had decided he wanted 'much wider coverage of street locations', and plans for additional permanent CCTV cameras at Hyde Park Corner, Oxford Circus, and Piccadilly Circus were well advanced. Police attention then turned to the eight cameras in the west London experiment, previously seen as purely a traffic control tool. Now that CCTV for public order had become a priority (that use itself having developed from traffic control), the west London traffic cameras took on a new use in general public surveillance, as the police started monitoring the camera feeds in their operations room the same year. The force's CCTV network for public surveillance therefore doubled in size overnight. What was more, 1970 saw the CITRAC plans starting to be noticed outside the traffic branch, too. For the first time, senior officers observed that the GLC scheme's first phase would soon add 45 CCTV cameras to the streets of the West End, a vastly greater number than that being planned under the 'official' public-order scheme.[84]

From 1971 onwards, the two schemes became intertwined. Plans were made with the GLC for the CITRAC cameras to be routinely available for public-order use once installed, and the public-order cameras being installed by the Metropolitan Police themselves were made available to the traffic control room in a reciprocal arrangement.[85] By 1975, the CITRAC cameras were beginning to come on line – just 14 at first – but dozens more soon followed, and by 1979 the entire scheme of some 150 cameras was complete and available to the public-order division as well as to the traffic cops.[86]

If the traffic and public-order discourses and camera networks intersected with each other behind the closed doors of New Scotland Yard and the GLC's County Hall, a similar obfuscation took place in public. Cameras within the traffic field had been well known since 1967, with Pathé newsreel films making clear that the streets were coming under the gaze of the eye in the sky, but *The Times* offered a cautionary narrative, noting the ability of the cameras to pick out number plates and the drivers' faces with their telephoto lenses, advising that 'in west London, Big Brother may soon be watching you'.[87] This manifestation of CCTV was certainly pretty hard to miss, and the muted public reaction to the west London cameras – few commentators appeared to pick up the *Times* line on Big Brother – gave the police confidence to increase their scope to public order three years later. Traffic became a fig-leaf. In a *Times* editorial in 1973, a police spokesman stated that the public-order cameras in Whitehall and Grosvenor Square – set up for the 1968 Vietnam protests – were in fact part of the CITRAC traffic scheme, a claim hotly denied by property holders whose buildings housed the cameras.[88] However, by 1979, this fiction had gone away. The CITRAC network was complete, parallel working between the public-order and traffic functions was routine, and the 'TV eye' was well known to the public for both its purposes.[89]

Traffic congestion had helped create a culture of remote surveillance that found new value in late-1960s fears of civil disorder. Congestion was the mechanism by which police control of political demonstrations transitioned from boots on the ground to the eye in the sky. This offers us a different route into the CCTV story of surveillance and crime reduction with which we are familiar: in the new reading the cameras were domesticated by their use in the benign social good of reducing congestion. But the cameras also played a crucial role in our wider account of electronics and sensor networks offering an alternative future for urban traffic – an alternative to the physical brutalism of planning. Alongside the loops in the road and the statistical models stored in the computer, the cameras demonstrated that cities could be modern and decongested without the cost and dislocation of destruction and construction. Area traffic control created new space for vehicles; the price was a requirement for all of us to submit to remote control, to accept remote surveillance, and to become statistics in the algorithms.

Conclusions

In the 1960s heist movie *The Italian Job*, Turin is brought to a standstill by Benny Hill's 'Professor Peach', who knocks out the CCTV network and loads a new program into the computer controlling the city's traffic light network. Michael Caine's 'Charlie Croker' moves in to capture $4 million from a bullion van caught up in the gridlock. 'Just think of it', he had mused as the plan was formed, 'A city in chaos. A smash-and-grab raid. And four million dollars through a traffic jam'.[90]

Writer Troy Kennedy Martin had bought the film's plotline from his brother, Ian, who had sold a London-based version to the BBC for a television play entitled *Ritual for the Steal*, which was never made.[91] Ian later recalled:

> Computerised traffic lights had just come into central London, so I put ... the big robbery ... in central London, with green lights all the way out and

creating traffic jams. I think the BBC felt that this would be too expensive to do, you know it was just ludicrously expensive to try and do this in London. So Troy bought these, it was two plays actually, and turned them into *The Italian Job*.[92]

The plot for one of the most iconic British films of modern times had come directly from the west London and CITRAC traffic light schemes – and the film was almost set in London. Computer control had reached cult status.

This chapter has examined an approach to traffic that appeared to offer a digital future more in harmony with public concern than the future offered by the postwar planners, with their brutally interventive road building. This was a new way of thinking, in which traffic was a system that could be controlled if it could be expressed in a mathematical model. Congestion was treated as a normal state that simply needed to be characterized in the algorithm. It existed inside the computer and could be reduced by the computer, rather than by concrete and construction. Unlike the civil engineers discussed in Chapter 3, these traffic specialists were not seeking super-efficiency, but simply least-worst delays. They wanted to usher a few more vehicles through each green phase in order to keep the diggers at bay. And if the diggers came, they were happy to offer their modelling skills to the about-to-be-built roads too. They were happiest working with virtual roads, but they were not divorced from the real world. They knew their models were complicated by hills and punctures and equipment failures. And they knew their models were political.

The 1960s catalysed debates about holistic traffic control that had begun much earlier, as we have seen in the discussions about police officers and highway engineers in the 1930s. Technology determined the new approach, to an extent. In a landmark 1968 House of Commons debate on commuting, at which Margaret Thatcher made an early contribution as shadow transport minister, the Minister of Transport, Richard Marsh, described a culture of technological urbanism as follows:

> No one form of transport – rail, bus or private car – can be seen in isolation from the whole … we should look at the hardware and the software available to our urban problem as a whole and see how we can use it. There is a wide range of hardware available to us at present – computerised traffic schemes, dial-a-bus, pedestrian aids, self-drive taxis, bus lanes and so on. A whole new field of traffic engineering is open to us.[93]

But systems thinking was also constructed by social and political conditions at large, whether the public unrest of 1968, or the decline of the postwar consensus, a story that we considered in Chapter 5. Whilst the technology offered a more holistic approach, the local London politics remained febrile. Area traffic control in London had begun as a Ministry of Transport project, supported by the RRL, but had been taken over by the GLC and the Metropolitan Police as it spread across the capital. For the police – who took sole charge of the CITRAC control room – this appeared to be the successful conclusion of a long process of empire-building. The new system was ostensibly a GLC-owned network, but by taking its control centre into the heart of New Scotland Yard, the Metropolitan Police was seeing off the challenge, with one senior police commander noting as follows:

> the more we can do to blur the sharp division of responsibilities, without conceding independence of action, the less likely that arguments in favour of a separate traffic enforcement agency will be expounded, at any rate by the GLC.

He went on as follows:

> we need the GLC's capital investment and scientific know-how in the design and provision of traffic surveillance equipment and they need our experience and man-power in dealing with abnormal congestion. Some sort of 'marriage' is inevitable and a joint traffic control operation centre should provide the setting not only for the ceremony but for a suitable honeymoon period thereafter.

He concluded that such a 'marriage' was vital 'if the police are to continue to have the dominant role'.[94]

However, it would be wrong to conclude that this was a sinister police takeover of a benign traffic technology in order to further an agenda of surveillance and control. The reality was more complex (as throughout this book) and more ambivalent. It took several years for the police to see CCTV as a desideratum outside the narrow confines of traffic. Perhaps it feared job losses if cameras could take over the work of foot-slogging constables, or perhaps it was just slow to move, lacking a technology champion such as Alker Tripp had been in the 1930s. A more nuanced conclusion is that network infrastructures take time to take root but, once embedded, reconfigure the options available to actors when external circumstances change, as happened in this instance with the popular unrest of 1968. A final comment is to observe that this computerized network is an *information* technology: it gathers and processes data about people's movement around London. If, as we saw in Chapter 5, road use can be marketized, then knowledge of the state of the market second by second could come to have great value.

Notes

1 *The Times*, 7 May 1971, 24.
2 *New Scientist*, 21 July 1977, 147.
3 See Simon Gunn, 'The Buchanan Report, Environment and the Problem of Traffic in 1960s Britain', *Twentieth Century British History* 22, no. 4 (2011): 521–42; and Simon Gunn, 'Introduction to *Traffic in Towns*', in *Traffic in Towns: A Study of the*

Long Term Problems of Traffic in Urban Areas (The Buchanan Report), ed. Colin Buchanan (Abingdon: Routledge, 2015), viii–xvi.

4 The National Archives (hereafter TNA), MT 111/64, 'Area Traffic Control by Digital Computer', draft of a speech by an unidentified Ministry of Transport official, n.d. but probably between 1963 and 1966.

5 J. Michael Thomson, 'The Value of Traffic Management', *Journal of Transport Economics and Policy* 2, no. 1 (January 1968): 5; Malcolm Buchanan, Nicholas Bursey, Kingsley Lewis, and Paul Mullen, *Transport Planning for Greater London* (Hampshire: Saxon House, 1980), Chapter 9; David Starkie, *The Motorway Age: Road and Traffic Policies in Post-War Britain* (Oxford: Pergamon Press, 1982), Chapter 3.

6 Daniel L. Gerlough, 'Control of Automotive Traffic on an Area Basis' (1954), unpublished technical memorandum cited in TNA, DSIR 12/92, 'A Review of Developments in Area Traffic Control', Road Research Laboratory, October 1961.

7 TNA, DSIR 70/6, 'Proposals for Research on Area Traffic Control', Road Research Laboratory, October 1960; TNA: DSIR 12/92, 'A Review of Developments in Area Traffic Control', Road Research Laboratory, October 1961.

8 TNA, DSIR 70/6, memorandum of interview, 20 March 1961.

9 TNA, DSIR 12/185, 'Road Research Board Panel on Area Traffic Control: Minutes of First Meeting', 16 October 1961, and 'Minutes of Second Meeting', 13 December 1961; detailed reports informing and arising from the panel's work are contained in TNA, DSIR 12/92, including descriptions of the worldwide state of the art in traffic control technologies in 1961.

10 TNA, DSIR 12/92, 'Panel on Area Traffic Control, Report to the Board', July 1962.

11 TNA, MT 106/405, 'Area Traffic Control in London', minute, unidentified author, n.d. but probably 1966; TNA, MT 111/59, 'West London Experiment: Tenth Report to the Management Committee on Behalf of the Working Party', June 1967.

12 TNA, MT 106/405, 'An Experiment in Area Traffic Control in Glasgow', RRL minute, May 1965; TNA, MT 111/110, press release issued by Plessey, 17 November 1967.

13 Richard Ham, 'Area Traffic Control in West London 2: Vehicle Counting Detectors', *Traffic Engineering and Control* 11, no. 4 (August 1969): 172.

14 TNA, DSIR 12/92, 'Proposal for an Investigation into Area Traffic Control by Co-ordination of Traffic Signals', Road Research Laboratory, February 1962.

15 TNA, MT 111/64, 'Computer-Control of Traffic in London', draft of an article for an Italian motoring journal, August 1968.

16 *The Times*, 19 September 1966, 8.

17 TNA, MT 111/59, reports of the Working Party to the Management Committee, 1964–8.

18 TNA, MT 111/110, press release issued by the Ministry of Transport, 22 January 1968; details of data transmission technologies in TNA, MT 111/64, 'Traffic and Computer Control', draft of an article for the *Journal of the Institution of Highway Engineers*, 1968.

19 One of the earliest practical demonstrations of a computerized vehicle-sensing network was given by William Vickrey to the US Congress in 1959; see William Vickrey, 'Evidence', in *Transportation Plan for the National Capital Region: Hearings before the Joint Committee on Washington Metropolitan Problems (86th Congress)*, ed. US Congress (Washington: US Congress, 1959), 454–585.

20 TNA, MT 111/64, 'West London's Computer Traffic Control System', draft of an article for an unidentified journal, n.d. This article offers substantial detail about the sensor network.

21 Ham.

22 Valerie E. Miller, 'Area Control by Digital Computer', *Traffic Engineering and Control* 5, no. 6 (October 1963): 365.

23 TNA, MT 111/64, 'West London's Computer Traffic Control System', draft of an article for an unidentified journal, n.d.

24 British Pathé, *Closed Circuit TV for Traffic* (film), 1967.

25 *The Times*, 24 January 1968, 9.

26 TNA, MT 111/60, 'West London Experiment Management Committee: Report to Minister', J. G. Smith, 19 July 1965; TNA, MT 111/64, 'Traffic Control', draft of a speech to the Institute of Transport by F. B. Green, November 1965.

27 TNA, MT 111/110, press release issued by the Ministry of Transport, 22 January 1968.

28 *The Times*, 6 May 1964, 11. On the Toronto system, see TNA, MT 106/405, 'Toronto's Digital Computer Controlled Signal System', Bureau of Public Roads, 8 April 1965.

29 *The Times*, 13 January 1967, 10.

30 TNA, MT 111/110, press release issued by the Ministry of Transport, 22 January 1968.

31 TNA, MT 111/45, 'Planning and Control of Systems for Road and Rail Traffic', symposium paper, H. A. Codd, Elliott Traffic Automation, 5 November 1963.

32 Frank Haight, 'Annotated Bibliography of Scientific Research in Road Traffic and Safety', *Operations Research* 12, no. 6 (1964): 976–1039.

33 Reuben Smeed, *The Traffic Problem in Towns* (Manchester: Manchester Statistical Society, 1961), 27.

34 Smeed, 36.

35 Smeed, 4.

36 Reuben J. Smeed, 'The Traffic Problem in Towns: A Review of Possible Long Term Solutions', *Town Planning Review* 35, no. 2 (July 1964): 133–58; Reuben Smeed, 'Traffic Studies and Urban Congestion', *Journal of Transport Economics and Policy* 2, no. 1 (January 1968): 33–70.

37 See for instance Michael Batty, 'Urban Modeling', in *International Encyclopedia of Human Geography*, ed. Nigel Thrift and Rob Kitchin (Oxford: Elsevier, 2009), 51–8; Anthony G. Wilson, 'Land-Use/Transport Interaction Models: Past and Future', *Journal of Transport Economics and Policy* 32, no. 1 (1998): 3–26; Richard Klosterman, 'An Introduction to the Literature on Large-Scale Urban Models', *Journal of the American Planning Association* 60, no. 1 (1994): 41–4; Stephen Philip Chichele Plowden, 'Transportation Studies Examined', *Journal of Transport Economics and Policy* 1, no. 1 (1967): 5–27; and Alan Voorhees, 'The Nature and Uses of Models in City Planning', *Journal of the American Institute of Planners* 25, no. 2 (1959): 57–60.

38 Ira Lowry, *A Model of Metropolis* (Santa Monica: Rand Corporation, 1964), 1.

39 Michael Batty, 'A Chronicle of Scientific Planning: The Anglo-American Modeling Experience', *Journal of the American Planning Association* 60, no. 1 (1994): 7.

40 David Harvey, *Social Justice and the City* (London: Edward Arnold, 1973); R. Andrew Sayer, 'A Critique of Urban Modelling: From Regional Science to Urban and Regional Political Economy', *Progress in Planning* 6, no. 3 (1976): 187–254; see also Mark Billinge, Derek Gregory, and Ron Martin, eds., *Recollections of a Revolution: Geography as Spatial Science* (London: Macmillan, 1984) esp. Part III.

41 For an overview, see Matti Pursula, 'Simulation of Traffic Systems: An Overview', *Journal of Geographic Information and Decision Analysis* 3, no. 1 (1999): 1–8; and E. T. Powner, M. Gill Hartley, Frederick George Heath, and David H. Green, 'Road Traffic Simulation Employing a Hardware Approach Philosophy and Design Considerations', *Simulation* 15, no. 3 (September 1970): 113–18.

42 Daniel Gerlough, 'Simulation of Freeway Traffic on a General-Purpose Discrete Variable Computer' (PhD thesis, University of California, 1955); cited in Pursula, 1.

43 *Science News Letter*, 19 February 1955, 120.

44 John Glen Wardrop, 'The Capacity of Roads', *Operations Research* 5, no. 1 (March 1954): 14–24; this computer simulation built on Wardrop's earlier work on the mathematical modelling of traffic, in John Glen Wardrop, 'Some Theoretical Aspects of Road Traffic Research', *Institution of Civil Engineers Proceedings, Engineering Division* 1, no. 3 (1952): 325–62.

45 See for instance David R. Cox, 'The Statistical Analysis of Congestion', *Journal of the Royal Statistical Society: Series A (General)* 118, no. 2 (1955): 324–35; and Frank Haight, 'Towards a Unified Theory of Road Traffic', *Operations Research* 6, no. 6 (1958): 813–26.

46 Richard Grimsdale, Ronald Mathers, and Frank Sumner, *An Investigation of Computer Controlled Traffic Signals (Unpublished Institution of Civil Engineers O.C. 6645)* (London: Institution of Civil Engineers, 1962); published in abstract as Richard Grimsdale, Ronald Mathers, and Frank Sumner, 'An Investigation of Computer-Controlled Traffic Signals by Simulation', *Proceedings of the Institution of Civil Engineers* 25, no. 2 (June 1963): 183–92.

47 TNA, MT 111/46, 'Simulation of Traffic in a Large Network of Signalized Intersections', by Daniel L. Gerlough and F. A. Wagner, paper prepared for presentation at the Second International Symposium on Theory of Road Traffic, sponsored by the Road Research Laboratory, 25–7 June 1963.

48 Miller, 361.

49 John A. Hillier, *The Glasgow Experiment: Schemes and Equipment*, RRL Report 95 (Crowthorne: Road Research Laboratory, 1967); Joyce Almond and Rex S. Lott, *The Glasgow Experiment: Implementation and Assessment*, RRL Report LR 142 (Crowthorne: Road Research Laboratory, 1968); John Hillier and Joyce Holroyd, 'The Glasgow Experiment in Area Traffic Control', *Traffic Engineering* 39, no. 1 (October 1968): 14–18; Michael Woolcock, *Traffic Signal Control in Glasgow by Computer*, RRL Report LR 262 (Crowthorne: Road Research Laboratory, 1969); Joyce Holroyd and John A. Hillier, *The Glasgow Experiment: PLIDENT and After*, RRL Report LR 384 (Crowthorne: Road Research Laboratory, 1971).

50 TNA, MT 111/64, 'Traffic and Computer Control', draft of an article for the *Journal of the Institution of Highway Engineers*, 1968; London Metropolitan Archives, City of London (hereafter LMA), LMA/GLC/TD/T/08/001, 'A System for the Monitoring and Control of Road Traffic', GLC draft report, May 1970.

51 Hillier, 2.

52 TNA, DSIR 70/6, correspondence between RRL and Ministry of Transport, 1964.

53 Hillier, 2.

54 Hillier and Holroyd, 14.

55 Holroyd and Hillier, 6.

56 Hannah B. Higgins, *The Grid Book* (Cambridge, MA: MIT Press, 2009), 49.

57 Gunn, 'The Buchanan Report', 528; see also Michael Hebbert, *London: More by Fortune than Design* (Chichester: John Wiley & Sons, 1998), 75–7.

58 Robert Bruce, *First Planning Report to the Highways and Planning Committee of the Corporation of the City of Glasgow* (Glasgow: Corporation of the City of Glasgow, 1945); Patrick Abercrombie, *Clyde Valley Regional Plan*, preliminary edn (Glasgow: Clyde Valley Regional Planning Advisory Committee, 1946); see also Roger Smith and Urlan Wannop, eds., *Strategic Planning in Action: The Impact of the Clyde Valley Regional Plan 1946–1982* (Aldershot: Gower, 1985), Chapter 6.

59 Scott & Wilson, Kirkpatrick & Partners, *Report on a Highway Plan for Glasgow* (Glasgow: Corporation of the City of Glasgow, 1965); Robert Hodgen and John J. Cullen, 'Recent Developments in Highway Planning in Glasgow', *Proceedings of the Institution of Civil Engineers* 41, no. 2 (October 1968): 223–45; John Cullen, *Glasgow Motorways: A History* (Glasgow: Craigmore, 2006).

60 Derek Wall, *Earth First! and the Anti-Roads Movement* (London: Routledge, 1999), Chapter 2.

61 Woolcock, 1.

62 Cullen, 11.

63 TNA, MT 111/64, 'Area Traffic Control by Digital Computer', draft of a speech by an unidentified Ministry of Transport official, n.d. but probably between 1963 and 1966.

64 TNA, MT 106/405, 'Area Traffic Control in London', minute, unidentified author, n.d. but probably 1966.

65 LMA, LMA/GLC/TD/T/08/005, 'Greater London Council CITRAC Area Traffic Control – Phase I: Description of Scheme for Grant in Principle', report, 8 August 1969; TNA, MT 111/70, 'Report (No. 2) of the Environmental Planning Committee', GLC, 10 May and 6 December 1971; Metropolitan Police Heritage Centre, 'Area Traffic Control in Greater London', Metropolitan Police Traffic Division, April 1974; LMA, LMA/GLC/TD/T/08/001, Marconi press release, n.d; LMA, LMA/GLC/TD/T/08/001, 'Computer Graphics for Area Traffic Control', GLC draft seminar paper, n.d. but probably 1971; *New Scientist*, 21 July 1977, 147–9.

66 D. I. Robertson, *TRANSYT: A Traffic Network Study Tool*, RRL Report LR 253 (Crowthorne: Road Research Laboratory, 1969); George Charlesworth, *A History of the Transport and Road Research Laboratory 1933–1983* (Hampshire: Gower, 1987), 153–4.

67 Robertson.

68 Jerry R. Peirce and Kristin Wood, *Bus TRANSYT: A User's Guide*, TRRL Supplementary Report 266 (Crowthorne: Transport and Road Research Laboratory, 1977).

69 TNA, MEPO 2/10898, 'GLC Traffic Restraint Policy: Effect upon Area Traffic Control', internal Metropolitan Police minute, 1 April 1974.

70 Metropolitan Police Heritage Centre, 'Area Traffic Control in Greater London', Metropolitan Police Traffic Division, April 1974, 17.

71 TNA, MT 111/59, 'West London Traffic Control Experiment Management Committee: Preliminary Report', September 1968.

72 Stephen J. Fay, 'Tough on Crime, Tough on Civil Liberties: Some Negative Aspects of Britain's Wholesale Adoption of CCTV Surveillance during the 1990s', *International Review of Law, Computers and Technology* 12, no. 2 (1998): 316; Nicholas Fyfe and Jon Bannister, '"The Eyes upon the Street": Closed-Circuit Television Surveillance and the City', in *Images of the Street: Planning, Identity and Control in Public Space*, ed. Nicholas Fyfe (London: Routledge, 1998), 257; Clive Norris and Gary Armstrong, *The Maximum Surveillance Society: The Rise of CCTV* (Oxford: Berg, 1999), 51–2; Clive Norris, Mike McCahill, and David Wood, 'Editorial. The Growth of CCTV: A Global Perspective on the International Diffusion of Video Surveillance in Publicly Accessible Space', *Surveillance and Society* 2, no. 2/3 (2004): 110–11; Benjamin Goold, *CCTV and Policing: Public Area Surveillance and Police Practices in Britain* (Oxford: Oxford University Press, 2004), 16; Benjamin Goold, Ian Loader, and Angélica Thumala, 'The Banality of Security: The Curious Case of Surveillance Cameras', *British Journal of Criminology* 53 (2013): 980.

73 Chris Williams, 'Police Surveillance and the Emergence of CCTV in the 1960s', in *CCTV*, ed. Martin Gill (Leicester: Perpetuity Press, 2003), 9–22; Sarah Manwaring-White, *The Policing Revolution: Police Technology, Democracy and Liberty in Britain* (Brighton: Harvester Press, 1983), 91.

74 BSSRS Technology of Political Control Group with RAMPET, *TechnoCop: New Police Technologies* (London: Free Association, 1985), 41–2.

75 TNA, MEPO 2/9956, report by Superintendent Preece, 1960 (emphasis added).

76 TNA, MEPO 2/9956, report by Superintendent Lacey, 1960 (emphasis added).

77 Williams, 13–14; for the press reaction to the royal visit, see clippings from the *Daily Telegraph*, the *Evening Standard* and the *Star* in TNA, MEPO 2/9956.
78 TNA, MEPO 2/9956, picture caption, July 1960.
79 Williams, 15.
80 TNA, MEPO 2/9956, report by Metropolitan Police Research and Planning branch on CCTV, 16 March 1965.
81 Williams, 16.
82 TNA, MEPO 25/7, list of forthcoming events requiring CCTV coverage by police, May 1969.
83 Williams, 17.
84 TNA, MEPO 25/7, notes of a meeting about CCTV, 9 June 1970.
85 TNA, MEPO 25/7, notes of a meeting about CCTV, 23 June 1971; note on Area Traffic Control, 24 September 1971; letter from Metropolitan Police to Home Office, 7 March 1972.
86 TNA, MEPO 2/10898, 'CCTV Cameras: Area Traffic Control', internal Metropolitan Police memorandum, 1 September 1975.
87 *The Times*, 13 January 1967, 10.
88 *The Times*, 31 January 1973, 14; see also Carol Ackroyd, Karen Margolis, Jonathan Rosenhead, and Tim Shallice, *The Technology of Political Control* (Harmondsworth: Penguin, 1977), 183–6.
89 *Guardian*, 26 January 1979, 5.
90 Paramount Pictures, *The Italian Job* (film), 1969. Thanks to Mark Swenarton for raising the significance of the film's plotline.
91 Lez Cooke, *Troy Kennedy Martin* (Manchester: Manchester University Press, 2007), 12.
92 Ian Kennedy Martin, interviewed on 19 March 2004, in Cooke, 45–6.
93 *Hansard*, HC Deb, 29 November 1968, Vol. 774, cols. 895–990.
94 TNA, MEPO 2/10898, internal Metropolitan Police memorandum, 29 April 1968.

Bibliography

London Metropolitan Archives: Greater London Council Papers.
Metropolitan Police Heritage Centre: Traffic Division Papers.
The National Archives: Department of Scientific and Industrial Research Papers.
The National Archives: Metropolitan Police Papers.
The National Archives: Ministry of Transport Papers.
Hansard.
Guardian.
New Scientist.
Science News Letter.
The Times.
Abercrombie, Patrick. *Clyde Valley Regional Plan*, preliminary edn. Glasgow: Clyde Valley Regional Planning Advisory Committee, 1946.
Ackroyd, Carol, Karen Margolis, Jonathan Rosenhead, and Tim Shallice. *The Technology of Political Control*. Harmondsworth: Penguin, 1977.
Almond, Joyce, and Rex S. Lott. *The Glasgow Experiment: Implementation and Assessment*. RRL Report LR 142. Crowthorne: Road Research Laboratory, 1968.
Batty, Michael. 'A Chronicle of Scientific Planning: The Anglo-American Modeling Experience'. *Journal of the American Planning Association* 60, no. 1 (1994): 7–16.
———. 'Urban Modeling'. In *International Encyclopedia of Human Geography*, ed. Nigel Thrift and Rob Kitchin, 51–8. Oxford: Elsevier, 2009.

Billinge, Mark, Derek Gregory, and Ron Martin, eds. *Recollections of a Revolution: Geography as Spatial Science*. London: Macmillan, 1984.

British Pathé. *Closed Circuit TV for Traffic* (film), 1967.

Bruce, Robert. *First Planning Report to the Highways and Planning Committee of the Corporation of the City of Glasgow*. Glasgow: Corporation of the City of Glasgow, 1945.

BSSRS Technology of Political Control Group with RAMPET. *TechnoCop: New Police Technologies*. London: Free Association, 1985.

Buchanan, Malcolm, Nicholas Bursey, Kingsley Lewis, and Paul Mullen. *Transport Planning for Greater London*. Hampshire: Saxon House, 1980.

Charlesworth, George. *A History of the Transport and Road Research Laboratory 1933-1983*. Hampshire: Gower, 1987.

Cooke, Lez. *Troy Kennedy Martin*. Manchester: Manchester University Press, 2007.

Cox, David R. 'The Statistical Analysis of Congestion'. *Journal of the Royal Statistical Society: Series A (General)* 118, no. 2 (1955): 324–35.

Cullen, John. *Glasgow Motorways: A History*. Glasgow: Craigmore, 2006.

Fay, Stephen J. 'Tough on Crime, Tough on Civil Liberties: Some Negative Aspects of Britain's Wholesale Adoption of CCTV Surveillance during the 1990s'. *International Review of Law, Computers and Technology* 12, no. 2 (1998): 315–47.

Fyfe, Nicholas, and Jon Bannister. '"The Eyes upon the Street": Closed-Circuit Television Surveillance and the City'. In *Images of the Street: Planning, Identity and Control in Public Space*, ed. Nicholas Fyfe, 254–67. London: Routledge, 1998.

Gerlough, Daniel. 'Simulation of Freeway Traffic on a General-Purpose Discrete Variable Computer'. PhD thesis, University of California, 1955.

Goold, Benjamin. *CCTV and Policing: Public Area Surveillance and Police Practices in Britain*. Oxford: Oxford University Press, 2004.

Goold, Benjamin, Ian Loader, and Angélica Thumala. 'The Banality of Security: The Curious Case of Surveillance Cameras'. *British Journal of Criminology* 53 (2013): 977–96.

Grimsdale, Richard, Ronald Mathers, and Frank Sumner. *An Investigation of Computer Controlled Traffic Signals (Unpublished Institution of Civil Engineers O.C. 6645)*. London: Institution of Civil Engineers, 1962.

———. 'An Investigation of Computer-Controlled Traffic Signals by Simulation'. *Proceedings of the Institution of Civil Engineers* 25, no. 2 (June 1963):183–92.

Gunn, Simon. 'The Buchanan Report, Environment and the Problem of Traffic in 1960s Britain'. *Twentieth Century British History* 22, no. 4, (2011): 521–42.

———. 'Introduction to *Traffic in Towns*'. In *Traffic in Towns: A Study of the Long Term Problems of Traffic in Urban Areas (The Buchanan Report)*, ed. Colin Buchanan, viii–xvi. Abingdon: Routledge, 2015.

Haight, Frank. 'Annotated Bibliography of Scientific Research in Road Traffic and Safety'. *Operations Research* 12, no. 6 (1964): 976–1039.

———. 'Towards a Unified Theory of Road Traffic'. *Operations Research* 6, no. 6 (1958): 813–26.

Ham, Richard. 'Area Traffic Control in West London 2: Vehicle Counting Detectors'. *Traffic Engineering and Control* 11, no. 4 (August 1969): 172–76.

Harvey, David. *Social Justice and the City*. London: Edward Arnold, 1973.

Hebbert, Michael. *London: More by Fortune than Design*. Chichester: John Wiley & Sons, 1998.

Higgins, Hannah B. *The Grid Book*. Cambridge, MA: MIT Press, 2009.

Hillier, John A. *The Glasgow Experiment: Schemes and Equipment.* RRL Report 95. Crowthorne: Road Research Laboratory, 1967.

Hillier, John, and Joyce Holroyd. 'The Glasgow Experiment in Area Traffic Control'. *Traffic Engineering* 39, no. 1 (October 1968): 14–18.

Hodgen, Robert, and John J. Cullen. 'Recent Developments in Highway Planning in Glasgow'. *Proceedings of the Institution of Civil Engineers* 41, no. 2 (October 1968): 223–45.

Holroyd, Joyce, and John A. Hillier. *The Glasgow Experiment: PLIDENT and After.* RRL Report LR 384. Crowthorne: Road Research Laboratory, 1971.

Klosterman, Richard. 'An Introduction to the Literature on Large-Scale Urban Models'. *Journal of the American Planning Association* 60, no. 1 (1994): 41–4.

Lowry, Ira. *A Model of Metropolis.* Santa Monica: Rand Corporation, 1964.

Manwaring-White, Sarah. *The Policing Revolution: Police Technology, Democracy and Liberty in Britain.* Brighton: Harvester Press, 1983.

Miller, Valerie E. 'Area Control by Digital Computer'. *Traffic Engineering and Control* 5, no. 6 (October 1963): 359–365.

Norris, Clive, and Gary Armstrong. *The Maximum Surveillance Society: The Rise of CCTV.* Oxford: Berg, 1999.

Norris, Clive, Mike McCahill, and David Wood. 'Editorial. The Growth of CCTV: A Global Perspective on the International Diffusion of Video Surveillance in Publicly Accessible Space'. *Surveillance and Society* 2, no. 2/3 (2004): 110–35.

Paramount Pictures. *The Italian Job* (Film), 1969.

Peirce, Jerry R., and Kristin Wood. *Bus TRANSYT: A User's Guide.* TRRL Supplementary Report 266. Crowthorne: Transport and Road Research Laboratory, 1977.

Plowden, Stephen Philip Chichele. 'Transportation Studies Examined'. *Journal of Transport Economics and Policy* 1, no. 1 (1967): 5–27.

Powner, E. T., M. Gill Hartley, Frederick George Heath, and David H. Green. 'Road Traffic Simulation Employing a Hardware Approach Philosophy and Design Considerations'. *Simulation* 15, no. 3 (September 1970): 113–18.

Pursula, Matti. 'Simulation of Traffic Systems: An Overview'. *Journal of Geographic Information and Decision Analysis* 3, no. 1 (1999): 1–8.

Robertson, Dennis I. *TRANSYT: A Traffic Network Study Tool.* RRL Report LR 253. Crowthorne: Road Research Laboratory, 1969.

Sayer, R. Andrew. 'A Critique of Urban Modelling: From Regional Science to Urban and Regional Political Economy'. *Progress in Planning* 6, no. 3 (1976): 187–254.

Scott & Wilson, Kirkpatrick & Partners. *Report on a Highway Plan for Glasgow.* Glasgow: Corporation of the City of Glasgow, 1965.

Smeed, Reuben. *The Traffic Problem in Towns.* Manchester: Manchester Statistical Society, 1961.

———. 'Traffic Studies and Urban Congestion'. *Journal of Transport Economics and Policy* 2, no. 1 (January 1968): 33–70.

Smeed, Reuben J. 'The Traffic Problem in Towns: A Review of Possible Long Term Solutions'. *Town Planning Review* 35, no. 2 (July 1964): 133–58.

Smith, Roger, and Urlan Wannop, eds. *Strategic Planning in Action: The Impact of the Clyde Valley Regional Plan 1946–1982.* Aldershot: Gower, 1985.

Starkie, David. *The Motorway Age: Road and Traffic Policies in Post-War Britain.* Oxford: Pergamon Press, 1982.

Thomson, J. Michael. 'The Value of Traffic Management'. *Journal of Transport Economics and Policy* 2, no. 1 (January 1968): 3–32.

Vickrey, William. 'Evidence'. In *Transportation Plan for the National Capital Region: Hearings before the Joint Committee on Washington Metropolitan Problems (86th Congress)*, ed. US Congress, 454–585. Washington: US Congress, 1959.

Voorhees, Alan. 'The Nature and Uses of Models in City Planning'. *Journal of the American Institute of Planners* 25, no. 2 (1959): 57–60.

Wall, Derek. *Earth First! and the Anti-Roads Movement*. London: Routledge, 1999.

Wardrop, John Glen. 'The Capacity of Roads'. *Operations Research* 5, no. 1 (March 1954): 14–24.

———. 'Some Theoretical Aspects of Road Traffic Research'. *Institution of Civil Engineers Proceedings, Engineering Division* 1, no. 3 (1952): 325–62.

Williams, Chris. 'Police Surveillance and the Emergence of CCTV in the 1960s'. In *CCTV*, ed. Martin Gill, 9–22. Leicester: Perpetuity Press, 2003.

Wilson, Anthony G. 'Land-Use/Transport Interaction Models: Past and Future'. *Journal of Transport Economics and Policy* 32, no. 1 (1998): 3–26.

Woolcock, Michael. *Traffic Signal Control in Glasgow by Computer*. RRL Report LR 262. Crowthorne: Road Research Laboratory, 1969.

7 The traffic problem and the mobilities of capital

Introduction

In Chapter 2, traffic in the history of the planning of London was surveyed. This raised a number of interrelated themes common to the dominant characterizations of the traffic problem, including ideas around governance, technologies, and segregation. Treating traffic as a socio-technical assemblage, this survey in turn offered encouragement to widen the network being examined, exploring these ideas in disciplines outside urban planning, as well as looking at proposals put forward to complement, or to oppose, the hard interventions favoured by the mainstream planning discipline. Chapters 3 to 6 offered case studies in response. One particular actor in the traffic network has hovered at the margins particularly of the accounts explored in Chapters 5 and 6 – namely capital – with both road pricing and electronic traffic management being framed in opposition to the capital-intensive road building programmes of the urban planners. It is necessary now to bring capital from the margins to the centre of the story in order to examine its role in the traffic network. To do so, this chapter will return the focus back to the professional planner, and will consider how aspects of the plans surveyed in Chapter 2 – particularly those of Patrick Abercrombie at the LCC, William Holford at the City of London, and Colin Buchanan at the Ministry of Transport – were turned into tangible reality on the streets of London after the war and the passing of the 1947 Planning Act. In doing so, the chapter will explore the role of traffic and planning policy in the relationships between capital and property.

No plan is a singular artefact or statement, but is a messy expression of countless ideas, all of them with histories and multiple authors, only some of which become reality through long processes of negotiations, and those negotiations ineluctably involved bargaining with private landowners, speculators, investors, and developers seeking to profit from property. As Stephen Elkin has remarked, 'development control was the principal means by which the [London County] Council attempted to serve its planning ends'.[1] This tense process means that the story of realizing traffic plans after 1947 involves people – politicians, planning officers, landowners, financiers, developers, road-users, residents and campaign groups – and capital. This chapter will demonstrate that long before the 1970s decline of the postwar consensus and the rise of overtly market-led approaches, plan-realization was a matter of negotiation in which traffic congestion had exchange value and created opportunities for the accumulation of capital.

The first location for study will be the junction of Tottenham Court Road and Oxford Street, known as St Giles' Circus, to look at the developer-led construction by the LCC of a traffic gyratory, part of a priority scheme in the 1951 LCC *Development Plan*. This will conclude that wider planning ends could be harmed if planners used developers to secure particular traffic projects – developers could extract planning permissions that were weighted strongly in their favour. Next, the focus will move to London Wall, a City Corporation-built dual carriageway opened in 1959, which was proposed in the Holden and Holford plan of 1947. A planner-led Corbusian techno-modernity of grids, vertical separation, and total reconstruction will be explored, concluding that here, planners drove a harder bargain than at St Giles' Circus, and used planning permissions in a more muscular attempt to see their traffic goals achieved, but that this, too, stumbled and fell, owing in large part to the shifting sands of property capital. Finally, a patchwork of residential streets in Pimlico will come into view. In 1967, this area was turned by Westminster Council into the 'Pimlico Precinct', London's first 'environmental area' following Buchanan's *Traffic in Towns* report. This project expressed a particularly complex relationship among traffic, property, and capital, and it will be concluded that traffic problems at a variety of scales offered opportunities that could be exploited for profit through the local planning system.

St Giles' Circus

Both the LCC traffic planners and the Ministry of Transport desperately wanted a roundabout at St Giles' Circus. This junction was a critical intersection where Tottenham Court Road, Charing Cross Road, Oxford Street, New Oxford Street, and St Giles' High Street met, and it had been on the council's priority list for re-engineering since the 1951 *Development Plan* placed it as London's sixth most congested junction.[2] The Ministry of Transport felt it was even more urgent than that.[3]

The problem was money. It was one thing to build roads in the outer reaches of London, where land prices and the densities of existing buildings were low. It was quite another to buy up prime central London properties for demolition to make way for tarmac. In April 1955, Kenneth Robinson, Labour MP for St Pancras North, led a Commons debate on London's roads. During it, he remarked on proposals to improve the St Giles' Circus junction, which was, he said, 'one of the worst congested intersections in London'. He used the junction daily and supported the plan. 'But these improvements are to cost nearly six million pounds and it is important to see what value we are to have for the money.'[4] This was the challenge facing the transport officials.

But there was a solution, and it came in the form of planning policy. In the mid-1950s, changes in planning legislation to favour private property development (including the abolition of the development charge), combined with easing of restrictions on building licences and materials, and a growing demand for office space in central London, saw the start of an office property development boom that was to last a decade – 'a decade of uncriticised and virtually untaxed profiteering', in Simon Jenkins's words.[5] Planning permission was the negotiating tool allowing London's councils to use the developers to provide the necessary land for road projects. The snag, as has been shown comprehensively by Jenkins, Oliver Marriott, and Stephen Elkin, among others, was that the developers held the whip hand, making fortunes out of property schemes that never quite delivered the traffic benefits promised, and promoting discrete road improvements that did not join up as the comprehensive plans intended.[6] And why should they? As will be seen, traffic congestion had value; without it, those fortunes might never be made.

The bargaining chip in the property boom was known as 'plot ratio', and it worked as follows.[7] All property developers wanted to build as much as possible on the land they acquired. This was how they made their money. Planning authorities, on the other hand, had wider concerns to take into account, and sought to restrict building density using the planning system. Before the Second World War, the tools they used under the London Building Acts were firstly a restriction on building height, and secondly the prescription of the angle from the opposite pavement to the top of the new building, which usually resulted in structures that stepped back from the road. After the war, a new system to control density, designed by William Holford, was introduced. Plot ratio was the relationship between the total amount of floor space within buildings erected on a development site and the actual land area of the site. It was prescribed as part of the planning process – 5:1 for most of the West End and City, for instance – and this gave the developers more options than under the old Building Acts.

Other restrictions aside, it meant that, for instance, in a 5:1 zone a five-storey building occupying all of the site was equivalent to a 20-storey building occupying only a quarter of the site, or any ingenious permutation in between the two extremes. The example of St Giles' Circus demonstrates the process. In 1966, a 34-storey skyscraper named Centre Point, designed by Richard Seifert's prolific practice, opened on the site. It was said by Oliver Marriott the following year that Centre Point was probably 'in absolute terms the most profitable single building ever promoted in this country'.[8] It came about because of the LCC planners and their wish for a congestion-busting roundabout. The story has been well told by Marriott, Jenkins, and Elkin, from whose detailed accounts the following summary has been drawn.[9]

In the early 1950s the LCC and the Ministry of Transport began negotiations to secure a ministry grant towards the cost of reconfiguring the St Giles' Circus intersection into a roundabout, widening certain roads to reduce the traffic congestion that had long dogged the junction. By 1956 a deal had been agreed, and the LCC approved plans to begin the work. The first step was for the council to purchase the properties on the area to be developed. One option open to the LCC was to declare the site a Comprehensive Development Area (CDA). This legislation enabled the council to make a compulsory purchase of land for large-scale redevelopment in areas of wartime damage, obsolete development, or bad layout, or to allow for the construction of an area defined in the *Development Plan*; and development could be carried out either by the council itself, a private firm, or both, so long as the plan itself was that of the LCC.[10] However, at this stage, St Giles' Circus was a simple roundabout scheme, rather than comprehensive development, and the LCC could not afford to buy more land than was needed for the road. This led the council to opt for an alternative: to use nineteenth-century legislation, which was normally far cheaper but which risked legal action on the part of the property owners, who would often press for greater compensation than the legislation allowed for. The LCC clearly hoped for a speedy resolution of any litigation, but, as has been observed, 1956 was right at the start of the property boom for central London office accommodation, and one key property owner on the site – herself a developer – was not willing to sell, so the case became mired in a lengthy Lands Tribunal.[11]

As the property litigation rumbled slowly on, the LCC began to have second thoughts about the scheme, moving towards a view that the road-only option would lead to a loss for planning, as the area surrounding the Circus would be ugly and the opportunity to improve it would be lost. Thus, in January 1957, the LCC entered into discussions with the Ministry of Housing and Local Government to see if it would support the council's declaring a CDA after all. It would not be impossible, the ministry responded, but would be risky, as the council's *Development Plan* would need to be altered, and the case might require a public inquiry, taking time and increasing costs. On balance, they advised against. This left the LCC in a bind. They felt it would be wrong to leave the wider area undeveloped, but they recognized the awkwardness of switching to the CDA option, having begun proceedings under the nineteenth-century Acts. One solution would be to work with a private developer, so, in March 1957, LCC planners put forward a wider scheme that could work either under a CDA or for private development.

Traffic congestion shines a light on the ambiguities of planning in London. The driving force behind the St Giles' Circus development was the LCC's desire to reduce congestion with a new gyratory intersection. To get it, the council was now considering allowing a private developer to build offices to finance the scheme. Yet at the same time, the LCC was seeking to *restrain* office development in central London, because it believed the office boom was increasing traffic congestion.[12] These were the same council planners, and the irony must not have been lost on them, but what else were they to do? Property in central London was colossally expensive, so they had to take seriously the option of getting into bed with property developers. The problem was that they were out of their depth. For Marriott, the LCC's planning department up to about 1955 was a 'team of brilliant creative minds' that were 'essentially academics' who 'displayed a mistaken naïveté about the developer'.[13] William Holford, author of the plot ratio system, preferred negotiating with architects than developers, because of 'a perceived mismatch of cultures', according to Bronwen Edwards and David Gilbert.[14] Jenkins felt that it was inevitable that the LCC's officers, 'trained in such conservative professions as architecture and quantity surveying[,] were not to be a match for the ingenuity of property developers and their allies'.[15]

Underlying the paradox of the potential offices-for-roads deal was the internal politics inherent in any organization. It is by now familiar that London's road schemes were at the mercy of warfare among ministries, borough councils, and the LCC (or GLC from 1966). St Giles' Circus was no exception. But the LCC itself was no homogeneous body; its engineers, planners, and accountants by no means always saw eye to eye. The proposal put forward by LCC planners restricted office space in the new development to more-or-less existing levels, with the central roundabout left clear of buildings. The engineers agreed. But accountants in the comptroller's department disagreed, proposing that a developer be allowed to build offices and shops to the maximum allowable under the *Development Plan* in order to maximize the return to the council from rents. The planning committee that considered the proposal recognized the tension but committed itself to comprehensive development of the area one way or another, unlocking the door for discussions with private developers. Matters rumbled on.

By 1959, it was clear that no money would be forthcoming from the LCC itself owing to financial pressures, which meant that the CDA option was off the table. Further pressure came from the report of the Ministry of Transport's Nugent Committee on London Roads, which revealed to Parliament the problems the LCC was having at St Giles' Circus.[16] And there was a new pressure. Litigation over the council's original property purchases was still ongoing, and new procedures were about to be introduced that would frown on the use of the nineteenth-century Acts for compulsory purchase, which did not allow a public hearing. The clock was ticking. Thus, in order to meet its commitment to comprehensive development, and to resolve its legal problems quickly, the planning committee now had only one option: to work with a developer, who could buy off the litigants at above-market rates, a tactic the council itself was not able to exercise.

In the meantime, word had got out into the property world, with a developer showing interest in autumn 1958. In July 1959, with the council now actively looking for a partner, the developer formally met with LCC officials and proposed to take on the scheme, but only if it could be progressed quickly. The developer was Harry Hyams, and the deal he placed on the table was simple. He would buy out the recalcitrant landowner as well as the rest of the sites needed for the development. He would then donate the land needed for the road intersection to the LCC. In return, he asked for a favourable planning permission for the rest of the site – and that would mean a tower block in the centre of the roundabout. This sort of deal was not at all unusual at the LCC; it had struck similar deals to fund traffic developments at Knightsbridge, Victoria, and Euston, as well as one at Piccadilly Circus, which, as Edwards and Gilbert have shown, ended up being abandoned.[17]

When it worked, the council got the land it needed without having to pay; the developer was allowed to develop the adjacent land for profit; and the whole deal was considered entirely proper. Why would developers give valuable land to the council for roads? The subtlety was in the plot ratio. The developer could transfer the plot ratio of the entire site onto the area left over for the office block. The developer therefore got the same lettable floor space; the downside, if one saw it as such, was that the building had a higher density than the planners had prescribed: essentially, taller tower blocks. Not only that, but when the LCC was so reliant on the developer for help in assembling the land, the developer had a strong hand in negotiations and would push for even higher plot ratio.

This tactic was described by the anonymous journalists of the Counter Information Services in their 1973 'Anti-Report' on property developers, which 'advised' budding developers that, when obtaining planning permission for an office development:

> Any problem with this can ... usually be solved by doing a deal with the council. These deals typically involve giving the council land for uses such as public housing or roads, in exchange for the necessary assistance with [Compulsory Purchase Orders] or satisfactory planning permission ... [I]t is ... wise to keep a job open in the office for any helpful, but underpaid, local authority official that you may run across.[18]

Hyams acted through his architect, Richard Seifert, whose first request was for a drawing showing the proposed route of the road network in the area. The LCC's engineers reluctantly put out a drawing, concerned that they were doing so blind to the developer's wider plans. Soon after, in late August 1959, both the LCC engineers and the developer's architect had second thoughts (or rather, they were working out how far they could push the deal). On Seifert's side, the proposed floor space was ratcheted up, in order, he claimed, to cover higher-than-anticipated land acquisition costs. He observed that this would nevertheless cause the LCC's legal challenges to disappear and that they would get their road at little cost. The LCC engineers, on the other hand, saw an opportunity to get wider carriageways in their roundabout now that the scheme was developer-led. Both sides came to agreement – an agreement that meant that a 30-storey block would occupy the roundabout, with a linked block to one side. In November this was agreed by the LCC planning committee.

More horse-trading ensued, with the engineers pressing for even more land-take for their road and the architect demanding an even taller tower and a higher proportion of its floor space given over to office use in return. This is where the internal politics at the LCC reveals where the power really lay, as all three key departments – engineering, planning, and finance – began to argue. Who won? Clearly, the traffic engineers and the accountants had the loudest voice on the council; the pleas of the planners to restrict office space and to keep the roundabout clear were overruled. The new proposal included the engineers' wider roads as well as a tower now reaching 34 storeys, a higher plot ratio and a higher proportion of offices than previously existed. April 1960 saw the final agreement and work began.

When Centre Point was completed at the centre of the roundabout in 1966, it rose to a towering 398 feet and its emergence from the ground had taken Londoners by surprise, such was the secretive nature of its planning permission (exceptions to this rule of secrecy tended to backfire against the developer, as Jack Cotton discovered to his cost at Piccadilly Circus).[19] The building lay unlet for many years – a profitable technique allegedly engineered by Hyams, though one that he denied until his death in 2015 – and eventually netted the reclusive tycoon an estimated profit of some £11.7 million.[20]

But what had been the public cost of this episode in postwar planning *Realpolitik*? In financial terms it was not so bad. It cost the LCC less to get its road scheme than it would have done even under the original minimum scheme, although a deal on ongoing rent payments went against the council. Yet in planning terms, as has been described, the ideals of the council had been ridden over roughshod. Office space in the area had significantly increased in direct opposition to LCC planning policy, and a tower block in a sensitive area had always been against the planners' wishes. Writing about the tensions of planning in the office boom in 1963, while it was still booming, Donald Foley claimed that 'Despite diverse and powerful political and developmental pressures, the London County Council acted with resourcefulness and firmness.'[21] With a decade of hindsight, Jenkins described deals such as St Giles' Circus as 'totally *ad hoc* ... concerned mainly with a vague desire to get development under way without impeding the traffic'.[22] A rueful planning officer explained the reality of such situations: 'it all hinged on how badly you wanted the road'.[23]

But there was a further irony. After all that, the roundabout was not even needed. Long before the new tower block was completed, in December 1960, the Ministry of Transport had come up with a different scheme to solve the congestion in St Giles' Circus: the extensive Charing Cross Road–Tottenham Court Road–Gower Street one-way system mentioned in Chapter 6, part of the new London Traffic Management Unit project. As Marriott concluded in 1967, 'the roundabout lay useless and unused'.[24] Jenkins was even more scathing in 1975: 'the developer, Harry Hyams, was permitted almost twice the normal plot ratio on his portion of the site in return for half a roundabout. This road has never been used.'[25] The LCC was furious, and objected to the Ministry of Transport, but could do nothing.[26] Across London at Euston, the LCC had granted permission for a huge office development in order to get an underpass built more cheaply, but this too showed who came out on top in the deal. 'The LCC was under the impression that the underpass would relieve traffic congestion', commented the Counter Information Services journalists. 'Ironically, it has had the opposite effect', they concluded.[27] Traffic congestion drove LCC planning decisions – inside the council, the engineers had the upper hand over the planners. But this just meant that the LCC could be played: by the Ministry of Transport, but more significantly by the property developers. Every actor was playing off the others to advance a position or make a profit. The traffic problem had value. But was it inevitable that property capital hold the whip hand over the planners? The case of London Wall, to which focus will now turn, involved a more muscular planning approach, though this too fell foul of the developers in the end.

London Wall

We met Michael Caine in the previous chapter, when we explored London's first computer-controlled traffic light system. There, we saw that the plot of Caine's 1969 film *The Italian Job* was drawn from the scriptwriter's experience of the real-life computerized experiment. Throughout this book, we have been charting alternative solutions to the traffic problem, and with this in mind it is easy to read *The Italian Job* as a comment on alternative modernities, with soft solutions such as traffic control coming off badly. The film presents us with a vision of a so-called high-tech smart city – in which traffic was controlled by electronic networks – failing spectacularly at the hands of Benny Hill's computer geek, Professor Peach. It also presented a fleeting glimpse of the hard solution. After his release from prison, Michael Caine's bullion-robber Charlie Croker hitches a ride on an early-morning milk-float, which passes along London Wall, a newly built dual carriageway to the north of the City of London. The road is almost deserted. Tower blocks rise on either side as Caine passes underneath a pedestrian walkway. There is no congestion, no traffic, no chaos. All is ordered. This, we are encouraged to feel, is the true modernity, the true technological vision. It is a modern planned city, unlike crowded, noisy Turin, which, even with the computer, still could not beat its congestion problem (or rather, the computer just made it worse).

The origins of the London Wall road project dated back to Holden and Holford's postwar report for the City of London, which designated a new 'Route 11' for the upper half of an inner ring road to ease congestion.[28] The bombing of the City of London in the Second World War had resulted in devastation across a large area to the north and north-east of St Paul's Cathedral and, using new powers granted by 1940s planning legislation, some 40 acres of land were bought by the City Corporation for redevelopment. The 1947 Planning Act gave planning control to the LCC (which was keen to ensure a coherent planned approach rather than piecemeal developments) and, in the ensuing years, plans were worked out between the two bodies for a four-part scheme.[29]

First came the modest Golden Lane residential estate, designed by Chamberlin, Powell, and Bon from 1952, based on slab blocks arranged on a controlling grid plan around a central tower block, with motor vehicles excluded under Alker Tripp's precinct or horizontal-segregation system. The second phase, planned from 1956 onwards and built through the 1960s and 1970s, was the much larger Barbican residential development by the same architects, again with slabs and towers laid onto a grid but with the addition of vertical segregation, with motor vehicles operating at street level and a two-storey pedestrian-only podium above. The pedestrian podium extended south to the third part of the scheme, a commercial development along the realization of part of the proposed Route 11 highway, renamed London Wall, which stretched from St Martin's Le-Grand in the west to Aldgate in the east. Plans were released in 1955 and comprised six commercial tower blocks and several smaller slab blocks, each arranged on a grid with pedestrian circulation at the podium level above the spine road below, which was opened in 1959 and dubbed 'the first new road to be built in the City of London for 88 years'.[30] The fourth aspect of the scheme, from 1959 onwards, was the extension of the raised pedestrian circulation network to the entire City of London, drawing from the conceptual framework for pedestrianism laid out in the Holden and Holford plan.[31]

London Wall was, of course, the conventional modernist techno-utopia that was surveyed in Chapter 2: the progressive vision of broad grid-plan highways and skyscrapers, vertical separation of pedestrians from vehicles, the linear city made manifest through total brutal reconstruction in concrete and steel. It was, as Nikolaus Pevsner observed, Le Corbusier's *Ville radieuse* realized.[32] Numerous 1960s films besides *The Italian Job* used London Wall as a location.[33] Back then it must have seemed like a planning dream come true.

At the time as the scheme was being constructed, commentators were gripped by its implications for urban planning. In 1963, Peter Hall observed that:

> the architect-planner no longer works in two dimensions, but in three; he has a series of blocks to fit into spaces, in an infinite number of surprising combinations. Not merely three levels are possible, as along London Wall, but six, seven, eight, all carrying separate functions.[34]

Edward Carter noted the rapid extension of the idea in London, stating that:

> the new way of segregation, by the complex interrelation of traffic and pedestrian routes on several levels, is now being applied by Chamberlin, Powell and Bon in their scheme for the Barbican; by the LCC architects in the latest part of the South Bank; and by Sir Leslie Martin in his scheme for the University Precinct.[35]

But one of the most complex critical analyses came in 1964 from Reyner Banham, who was working up ideas about the relationships among people, architecture, and technologies that were to be published a few years later.[36] In the 1964 BBC documentary film *A City Crowned with Green*, Banham urged the repopulation of central London under modern conditions, in compact high-rise flats with elevated segregated circulation for pedestrians and the demotion of motor traffic to the existing streets.[37] He introduced his idea as follows:

> In the old London and in the new, the vital thing is always the handling of the pedestrians. If there is something that is worth fighting for in the heart of London, something that is worth keeping, even when we knock down the

buildings, it is the network of pedestrian communications and squares – to stroll in, to sit in, just stand about in.

For Banham, pedestrianism was the basis of European civilization, but it was under threat from modernity:

> If we're going to keep it, then our urban renewal experts are going to have to find a way of keeping it *alive*, even under the apparently hostile conditions of motorized traffic systems. The best offer we have to date is to keep it not under that system but literally above it.

The camera tracked along the completed office tower blocks and unfinished building sites along London Wall. Banham observed the following:

> The high-level pedestrian terraces and the subterranean car parks along London Wall are our most extensive effort to date in dealing with the growing problems posed by the crush of cars, and in creating a complete second city for foot passengers, raised a couple of storeys above the world of wheels; made independent of the street pattern below, by providing bridges, and other connections, that make it never necessary to come down to street level and fight your way through the traffic. The idea's not new. It's been proposed before in London. But it just hasn't caught on.

The irony in Banham's footage of the London Wall development was that it depicted the majority of pedestrians down at street level, rather than using the pedestrian walkways. Partly this was simply owing to the fact that the elevated network was not complete. But, over the ensuing years, the 'pedway' scheme (as the walkways became known) failed to live up to its planners' hopes, as Michael Hebbert has shown.[38]

The pedway experiment came about through the unusual planning regime under which the London Wall commercial development operated, driven by the LCC's involvement and its fears of a piecemeal, developer-led assemblage of buildings.[39] This planning regime was quite different from the more submissive approach operating at St Giles' Circus and elsewhere. Under the leadership of the City's planning officer, H. Anthony Mealand, and the LCC's architect, Leslie Martin, the modular layout, heights, and bulks of the sequence of the six towers and numerous slabs were strictly defined before plots were sold off to commercial developers for design and construction. Crucially for the present story, the planning permissions also contained strict requirements for developers to include integrated pedestrian walkways and podium decks, and connections to the Barbican estate and across London Wall. Hebbert connects this to two movements that were considered in Chapter 2: the first being the idea of vertical segregation through multi-level development, which was strongly held by architects and planners at the LCC; the second was large-scale postwar urban redevelopment, which took the idea to heart.[40] To these movements can be added a third: the office development boom that was considered in the previous section.

The London Wall and Barbican scheme gave the LCC planners their chance to realize technological utopias of vertical segregation. Urban reconstruction provided the political traction, but in this model, it was the commercial developers who were expected to pay for and build the planners' pedway network. It started out well. In 1957–60, developers (including Harry Hyams) were quick to buy up the London Wall tower-block plots, with their commitments for podium decks and walkways.[41] The six towers were built as planned, 'not one of which yielded its lucky developer less than a million pounds' profit', stated Simon Jenkins.[42] Things were moving quickly and the office boom was at its height, so the scheme was expanded. The City Corporation, by then as committed to modernism and segregation as the LCC architects, decided to incorporate provision for pedestrian walkways into *all* new commercial developments in the Square Mile and began enacting this new policy through the planning consent system, sure that in the property boom a 30-mile network would emerge quickly. Figure 7.1 shows the network as proposed (in secret) in 1963.[43]

And the network did indeed start to emerge through the 1960s, as major new developments incorporated the prescribed provision for upper-level entrances and reception halls, with abutments for connecting bridges. This was all part of the 1960s world of negotiation, of extracting commercial benefit from traffic plans, but it seemed to be going more in the planners' favour than at St Giles' Circus. London Wall, based on a vast City-owned reconstruction site, strengthened the

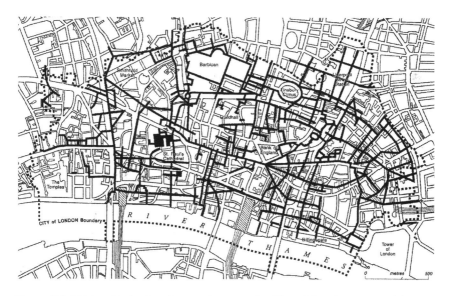

Figure 7.1 Thirty-mile 'pedway' network across the City of London proposed in 1963 (courtesy of Michael Hebbert).

planners' resolve, with the office boom apparently enabling the model to be applied on a widespread basis without crippling public cost. As Hebbert notes:

> The City of London seemed to have achieved the benefits of comprehensive redevelopment without the expense and intervention of public land assembly. Developers, for their part, were happy enough to include upper-level circulation in exchange for plot ratio concessions.[44]

In 1969 the GLC took on similar planning powers to create pedway networks elsewhere in London for eventual connection to those of the City, working with Westminster Council on bold plans for a comprehensive segregated network for pedestrian movement.[45]

But the realization of a techno-utopian segregated city was as short-lived as the office boom, which was rapidly cooling in the early 1960s and pretty much bust by the middle of the decade. As early as 1966, the City of London was acknowledging a 'failure of planning' owing to 'lack of office development permits', leading to an abundance of 'partly completed re-development units' as well as 'pedestrian walkways going from nowhere to nowhere', in the words of its planning committee chairman.[46] By the early 1970s, as an economic recession was starting to take hold, the vertical utopia looked increasingly costly to build and maintain, as well as simply unwanted by its intended users. As we saw in the Banham documentary, the public generally preferred street level. 'Why reproduce a ground-level network that, despite bombing and reconstruction, was still largely intact and intensively used', asks Hebbert.[47] There was simply no pay-off. Pedestrians did not want to use the pedways, developers resented the space they took up, councils increasingly had to pay for their upkeep, and the model of creating a network out of many independent commercial developments had broken down.

The wider modernist planning dream that was turning sour, too. As far as the residential Barbican was concerned, the cracks started to appear before the development had even been completed. David Heathcote observes that from the early 1970s, 'the majority of comments about the project were negative'. He cites a 1973 *Architectural Review* editorial that pointed to 'that coarsening of the visible man-made world which the Corbusian reality has brought'.[48] The following year, in a *New Society* article, Reyner Banham expressed what Heathcote sees as 'the sense of disappointment and betrayal felt by [him] in what he saw as a reversal of the purpose of the Modern architecture that he had championed for so long'. Heathcote considers that by this time, 'the negative moment of Postmodernism was upon us'.[49]

In the mid-1970s, the City Corporation underwent a radical policy-shift towards conservation. This did not happen in isolation; elsewhere in London, politicians and the public alike had spent a decade or more waking up to the effects of comprehensive redevelopment on the capital's historic fabric, with the death of the Ringways, as discussed in Chapter 2, being closely followed by the cancellation of a vast Covent Garden redevelopment project.[50] A turning point had been reached in public opinion towards comprehensive planning and, in the City, opportunities for redevelopment slowed. The comprehensive pedway network therefore failed to materialize.

Of course, the property market bounced back in due course. The pedways might have re-emerged. But the politics of capital and planning had changed. From 1986, Margaret Thatcher's financial deregulation project known as the Big Bang, with its changes in trading technologies and requirements for new office building layouts, combined with threats from the City's fast-growing younger sibling in Docklands, meant that conservation was all but abandoned. A huge building programme began again – but this one was market-led, not planner-led.[51] An early casualty was London Wall's planned purity, where a new air-rights development, Terry Farrell's postmodern Alban Gate office building, completed in 1992, straddled the roadway, unlocking the rampant redevelopment of the area in the years that followed.[52]

The reasons for the decline and fall of the pedway network were complex. But capital played a crucial role. Property developers always looked to exploit the planning system to make money. They were prepared to build in traffic infrastructure such as pedways when the market was booming, but were increasingly reluctant at times of bust. The planners had been bold in their ambitions and their careful use of the planning permission system to extract planning ends out of the developers. But the organic growth of the pedway network meant it did not happen quickly enough to ride the wave of the market and, when boom returned, the relationship between planning and capital had changed to a market-led approach. The pedway policy was abandoned and so, too, were the City's ideas of Corbusian grids, segregation, and tight modern planning control that had seen such a short blossoming.[53]

Most significantly for traffic congestion, Route 11 was never extended beyond the short stretch built in the 1950s, as the zeal for major urban road construction in London's heavily built-up areas was tempered and then crushed by public opposition and the rise of conservationism and environmentalism. As early as 1967, Oliver Marriott had described Route 11 as 'a fairly unimportant link' and 'one of the more absurd stretches of urban planning ... a six-lane highway which appears to go from nowhere to nowhere'.[54] In the end, despite Michael Caine and Benny Hill's best efforts to prove the opposite, it was the computerized traffic light network, rather than the *Ville radieuse*, that was extended into the twenty-first century. But the planning vision was not just about thoroughfares, tower blocks, and walkways in the sky. From Tripp and Abercrombie to Buchanan, the local neighbourhood also came under scrutiny. The developers had their eyes on this too.

The Pimlico Precinct

Lupus Street runs up from the northern bank of the River Thames, by the railway bridge into Victoria station, before turning north-eastwards towards today's

Pimlico tube station and Vauxhall Bridge. Along its southern side is Churchill Gardens, the pioneering postwar Powell and Moya council housing estate, as well as the notorious 1930s flats at Dolphin Square, home of establishment scandals that still make the headlines today. The houses and flats to the north of Lupus Street are more modest. The area was laid out by Thomas Cubitt in the 1830s for the duke of Westminster as part of his St George's Estate.[55] Cubitt's layout took the form of a grid that, as has been noted in this account, was alien to London at a large scale but fairly common in local areas, as here.

This part of the estate was completed after Cubitt's death in 1855, and is a neighbourhood of largely three-storey, stucco-fronted town-house terraces. By the 1960s, though, this quiet residential neighbourhood had become a series of vehicle rat-runs. As has been repeatedly observed, grid patterns reduce congestion in urban areas as drivers have numerous options to get from one point to another, even if there are blockages. This was why traffic experts from cops and engineers to modellers and planners yearned for a grid to replace London's muddle of streets. As traffic grew in the 1950s and 1960s, this 30-acre corner of Pimlico had become the haunt of motorists seeking short-cuts from the south-west along the Thames Embankment into Parliament Square and Victoria. This led to accidents on the residential streets, particularly at junctions, and experts felt the problem would only increase. In 1965, Westminster City Council decided to act. It felt the area's 5,500 residents deserved their peace and quiet – their amenity – so it formed a scheme to exclude through-traffic from these streets, turning the area into the 'Pimlico Precinct'.[56]

The challenge facing the council's chief engineer, Francis Cave, was how to exclude through-traffic while allowing residents free access to drive to and from their homes. His solution was simple. By a combination of one-way streets, selective closures, and banned turns, he was able to restrict entrance points into the precinct to just four streets, and no vehicle journey could be made through the area without making at least two turns – in other words, it became impossible to drive through the precinct in a straight line. Figure 7.2 highlights the route options open to motorists under the scheme.[57]

Moreover, footpaths were widened at junctions – Cave termed them 'blips' – to reduce vehicle speed and aid pedestrian crossing, and roads were limited to one lane in each direction, additional road width being turned into parking bays.[58] Trees were planted and seats installed at the newly widened intersections. Through-traffic would find it so time-consuming and complex to drive through the precinct that the benefit would be lost and drivers would return to the thoroughfares surrounding the area, which had enough capacity to take the marginal extra traffic. Accidents would be reduced and this corner of Cubitt's London would be returned to its peaceful state. The £39,000 scheme was approved for construction by the council in February 1966 and put into operation in April 1967 following the usual financial wrangling with the LCC and Ministry of Transport.[59]

The Pimlico Precinct was the first scheme of its kind to be installed in an existing neighbourhood rather than a redevelopment area, and was hailed by Westminster Council as a great success. Traffic levels across the precinct were

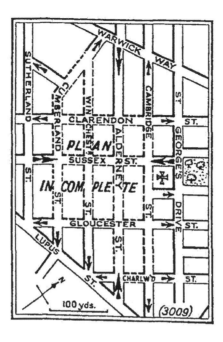

Figure 7.2 Route options under the Pimlico Precinct scheme, published in 1966 (Westminster City Council, courtesy of City of Westminster Archives Centre).

down, one survey location recording a drop from 260 vehicles per hour to 5. Accidents were down, from 11 to 1 per year, with no increase in accidents on the perimeter thoroughfares. And the public's reaction, Cave stated, was overwhelmingly positive – 'residents are delighted with the improved "atmosphere" in which they live and have taken the trouble to say so'.[60] Of course it was modest, but then, piecemeal improvement of single streets, junctions, or little pockets of the capital was, as has been noted throughout this study, the reality of traffic planning throughout the twentieth century. This one, however, seemed as though it could be scaled up. Cave commented that:

> it has been suggested that the Pimlico Precinct Scheme may have added 30 years' life to the buildings within its boundaries. The City Council has already resolved to extend the schemes both westward and eastward … The future road pattern of Westminster may well be based upon a number of similar precincts between which the district distributor roads are located.[61]

Cave's description of precincts divided by distributor roads is, of course, straight from Colin Buchanan's *Traffic in Towns* report published in 1963, and the Pimlico Precinct was London's first manifestation of Buchanan's idea of 'environmental areas'. As Carmen Hass-Klau has demonstrated, British ideas about urban neighbourhoods with restrictions on traffic went back at least as far as the Port Sunlight scheme of the 1880s and were clearly visible in Welwyn Garden City, designed in 1920.[62] Similar ideas were developed in Germany across this period and beyond.[63] In the USA, the suburb of Radburn, in New Jersey, was founded in 1929 with a pioneering street layout separating pedestrians from motor vehicles that was a strong influence on town planners subsequently. The Radburn layout was created specifically in opposition to the dominant grid layout of US cities, with their negative effects for pedestrians noted in Pimlico 30 years later.[64] Here we have a reversal of the aforementioned planners' desire to superimpose a grid – whether concrete or electronic – onto the non-grid of London's streets. The environmental area idea, with its long history, was an attempt to de-grid the grid (visions of the perfect plan were always ambivalent and contested).

Ideas from Radburn, themselves influenced by earlier British and German practice, were imported back into the UK by planners such as Raymond Unwin, whose official government reports of 1929 and 1933 were considered in Chapter 2. As was also shown in that chapter, Alker Tripp's advocacy for neighbourhood precincts had a great influence on British town planning from Abercrombie to Buchanan, who renamed them his environmental areas.[65] But concrete schemes for turning the idea into reality in London were slow to catch on. As Hass-Klau remarks, 'The astonishing story was how little was made out of it', although perhaps not so astonishing given the government's clear policy at that time of expansion of motor-car use. Furthermore, she notes the increasing use in the traffic planning profession in this period of computer modelling, as discussed in Chapter 6, whose practitioners were:

> technically skilled ... but they had lost, maybe unconsciously, for the first time most of their social objectives, and had little in common with the visionary planners of the turn of the century. Their counterparts, the traffic engineers, were more concerned with building roads, and details like environmental areas could be considered once all the roads were built.[66]

Yet, by 1973, some 150 environmental areas were planned or in existence across the UK, which elevates the 1965-designated Pimlico Precinct, as London's first, to a scheme more significant than its size might otherwise have suggested.[67] As Cave explained, Pimlico was simply a pilot scheme for more general application throughout Westminster once the principle had been proven.[68] This could be seen as a win for the environment. However, the other two case studies examined in this chapter have revealed the hand of capital in postwar urban planning and traffic, focusing on property speculation and the succession of property booms from the 1950s onwards. With that in mind, it would be prudent to ask why Westminster Council decided to adopt the environmental area with such apparent alacrity so soon after its promotion in *Traffic in Towns*. One can then ask, why Pimlico? And who were the residents in 1969 who told Westminster Council that they were 'delighted' with the 'atmosphere' of the area following the installation of the scheme?

The rat-running traffic through these side streets was itself a consequence of congestion-relief measures taken years before. Decentralization and the interwar growth of the suburbs had created the daily tide of commuters flowing inwards each morning and outwards at the end of the working day. Originally built as good nineteenth-century housing for the middle classes, the 42-acre St George's Estate had 'gone to seed', in Oliver Marriott's words, after troops were billeted there during the First World War.[69] The landowner, the duke of Westminster's Grosvenor Estate, had little control over the properties let on long leases, and, in a surprise move, disposed of the entire holding in 1950.[70] But some could see an opportunity in the seedy district. The growth of the suburbs was generating opportunities in the central London properties left behind, as those who were sick of the daily grind of the commute were beginning to eye up the centre of town again. For Reyner Banham, as was seen earlier in this chapter, this repopulation of the centre would be done on hard, segregated modern lines, with pedestrian networks superimposed over the streets. The reality was softer, more modest, and more conservative.

With the establishment hand of the duke of Westminster gone from the housing north of Lupus Street, the field was open for the developers to move in and, over the years following the 1950 sale, the estate passed through the hands of a variety of property developers, its price rising at each sale, before ending up in the hands of the notorious speculator Max Joseph in 1959, who sold out to developer Kennedy Leigh in 1962. Kennedy Leigh's son Gerald, a director of the firm, bought Hampton, the estate agent, in 1965, just as Kennedy Leigh went into liquidation and the St George's Estate was sold again – Hampton's managing the sale with an advertisement noting 'redevelopment potential'.[71] Property was a cosy, if cut-throat, world. After the 1965 liquidation sale, the estate ended up in the hands of the Royal Liver Assurance Company of Liverpool, which had become associated with Joseph a few years earlier in a Paddington development.[72] In early 1967 – just about when the Pimlico Precinct scheme went live – Royal Liver sold the headlease to developers Central Estates (Belgravia).[73]

Simon Jenkins said that this succession of developers had 'turned Pimlico into a property speculators' paradise and a residents' nightmare', while the *Evening News* commented acidly that 'somebody is making a great deal of money out of Pimlico. Property is changing hands at vast profits to the developers, and in the process ordinary people are being squeezed and their lives changed.'[74] What was happening at Pimlico was gentrification – long-term working-class tenants were, it was alleged, being systematically harassed and evicted by the new headlease holders and their agents, in collusion with the big institutional freeholders such as Royal Liver, so that the properties could be turned into flats for sale on the open market to middle-class owner-occupiers relocating from the suburbs.[75]

For Jenkins, it was not just the developers and financiers who were to blame for this insidious practice. 'For all the ingenuity of the property industry in these years', he remarked, 'nothing should detract from the crucial role played in many of the biggest deals by public authorities, in particular the London County Council'. Every major property deal in this period required the LCC's approval, 'and many were the direct result of its active participation', Jenkins observed, going on to note that LCC planners were 'confronted by immense pressure from the property industry'.[76] With such a complex network of actors and conflicts it is impossible to say beyond doubt that Westminster Council came under pressure from the LCC or Pimlico's parade of property developers to improve the amenity and safety of the St George's Estate – and therefore to increase its value – by creating the Pimlico Precinct. There were simply too many shareholdings spread across all sorts of shell and holding companies, and too many institutions involved in any deal. Clearly, the property industry had close links with the establishment and government, with financial interests spread far and wide.

What can be said is that the community of residents polled by City Engineer Francis Cave in around 1969 was not the same as those living in the area 15 years previously. In about 1973, the urban theorist Donald Appleyard visited the Pimlico Precinct to interview residents, and it was clear to him that the area had undergone high population turnover owing to 'a wholesale conversion of the community to high-priced apartments'. He claimed that this was due to the systematic policy of Westminster Council to approve planning applications by developers to convert homes from family rented units to retail studio and one-bedroom flats. They had the option to withhold permissions in order to retain low-income family homes in Pimlico but refused to exercise it. 'The traffic scheme may well have provided a catalyst for this trend', he noted, 'although it cannot be blamed as the cause'. Nevertheless, he observed, 'the close connection between traffic management and housing change cannot be ignored'. He concluded by saying 'Whether the original traffic engineers were aware of the possible social consequences of the traffic management scheme, we do not know.'[77] Aware or not, their actions here and elsewhere were part of a complex relationship among traffic, politics, planning, property, and profit.

Conclusions

In 1954, the outgoing Commissioner of the Metropolitan Police, Harold Scott, wrote about traffic congestion in the capital as follows: 'It is not plans, but a decision to put plans into operation that is needed ... The alternative is a creeping paralysis which in the end must prove fatal.'[78] But this suggests a binary situation – plan or no-plan – and also it assumes the planner is the only agent in the decision. Here we have looked at other agencies, particularly those who pay and those who use, as well as alternative visions of the future shape of London, especially the commercial viewpoint of postwar property speculators such as Harry Hyams, whose 'dynamic individualism' was seen by Frank Mort as one of a set of 'foils to Forshaw and Abercrombie's high-minded, ethical vision'.[79]

Mort's approach to the nature of plans – as cultural phenomena, performed by specialists and the public alike – is valuable in defusing the heroic planner account and helping us think about planning (and planners and plans) in different ways. The case studies in this chapter draw from Mort's approach. Plans have been seen to work in different ways from those intended by their nominal authors. They have been deployed within processes of marketization, in which traffic and its physical configurations (whether built into the streets or temporary congestions and flows) have been seen to have exchange value in the intense capitalization of London's urban space since the Second World War. Plans, planners, and planning have created a diverse range of new opportunities for people to make profits.

The architecture critic Ian Nairn wrote an *Observer* piece in December 1966 that captured the mood of the moment. Centre Point had just been completed – the view of it looming over Denmark Street's Tin Pan Alley was 'the kind of shock contrast that makes London unique'. But the office boom was over, giving London a 'lull in the building frenzy' that was 'a rare opportunity, here and now, for saving London'. The pedestrian network covering the Cities of London and Westminster was struggling to be realized, and Nairn proposed it be abandoned, claiming that 'the elaborate two-level system … would wreck the existing environment at enormous cost, in order to support a volume of traffic which Central London doesn't need'. More to his taste was the Pimlico Precinct:

> define the limits of each of London's villages, and go to work on the things – often small and inexpensive things – which make it more pleasant. If Pimlico needs to gain identity by stopping up some of the interminable intersections … then do so.

The ultimate problem, he observed, was the convoluted planning system, with different planning teams within and between different agencies competing with each other, and all of them ignoring the general public.[80] This was the most obvious change that took place in the late 1960s and early 1970s – the turn to the environment and to the people.

But this shift from planner-knows-best to planning-by-referendum – this degradation of the power of the professional town planner – took place alongside a more insidious change. That change, as this chapter has charted, was the rise in the power of the postwar property industry and the ability of landowners, developers, and capital investment funds to change London's streetscape by influencing the planning process through its many loopholes and unintended consequences. The acclaimed analyst Anthony Sampson, writing in the thick of the 1960s office boom, suggested that 'the very jumble and messiness of London, its zigzags, curves and old buildings, yield great prizes for those who can disentangle it', meaning insurance companies and property developers.[81] Streets and their users, as well as buildings and their inhabitants, had become the currency of a new form of property capitalism that sought to trade as much as invest, and to do so over the short term rather than the natural long-term cycles of bricks, mortar, concrete, and tarmac.

How were the grand plans of Abercrombie, Holford, Buchanan, *et al.* to be built? The answer, of course, came down to money, but in the new short-term world of trading for quick profits, with council planners struggling to keep pace with the loopholes of their own legislation, that meant that fragments – a roundabout here, a short stretch of highway there – were all that could result. Those who held the keys to capital took advantage of any opportunity to turn a relatively quick profit. This was most clear when the traffic problem was congestion, such as at St Giles' Circus. Developers could leverage land-for-planning concessions with the results, such as Centre Point, making them millionaires. Some planning regimes were more muscular in their relationship with capitalism than others, with the City of London extracting strong concessions from developers in return for permissions at London Wall and elsewhere, but here the long timescales of planning turned out to be incompatible with the short boom-and-bust timescales of property capital, and planning failed here, too, though for a complex set of reasons.

Less obvious, though, has been the ability of property capitalists to make money out of much smaller scales of traffic project, which exist only at the margins of the literature and the built environment of London. The Pimlico Precinct was entirely low-key. It attracted little attention, and can be easily missed today, comprising trees, benches, and one-way signs, not concrete tower blocks and elevated walkways. It hardly looks like a product of the twentieth-century modern planning vision. But it was not a project from the post-Ringways world of the environment and the humbling of the motor city. It was a mid-1960s project contemporary with the high-water mark of the car in the city. Such projects showed that money could be made, even if the traffic problem was *not* congestion but rather its inverse, namely free-flowing traffic in an efficient grid pattern of streets. The idea was Buchanan's (or even Tripp's), but its implementation was carried out by local councils that had close relationships with property developers. Congestion could be engineered *in* by traffic-calming, pushing up property evictions, conversions, prices, and gentrification.

Come boom or bust, whatever the planners and officials came up with to solve the traffic problem, the capital's property tycoons have found ways to make money from it. They have understood the value of London's streets and the profit that could be made from the public's need to use them. And, given the complex interrelationships among the different actors in planning, including politicians, officials, funders, developers, owners, and the wide range of other specialists including cops, engineers, scientists, and economists, there were many interests in keeping the situation complicated and the problem unsolved. In fact, it could be concluded that congestion was good for everybody except the public – no, even the public, who held their pensions and insurance policies with the big investment funds and therefore held stakes themselves. Everybody had an interest in congestion.

Notes

1 Stephen Elkin, *Politics and Land Use Planning* (Cambridge: Cambridge University Press, 1974), 26.
2 Elkin, 56.
3 Ministry of Transport, *London Traffic Congestion: Report by the London and Home Counties Traffic Advisory Committee* (London: HMSO, 1951), 16.
4 *Hansard*, HC Deb, 7 April 1955, Vol. 539, cols. 1331–55.
5 Simon Jenkins, *Landlords to London: The Story of a Capital and Its Growth* (London: Constable, 1975), 215.
6 Oliver Marriott, *The Property Boom* (London: Pan, 1967); Elkin; Jenkins.
7 Marriott, 43–4.
8 Marriott, 135.
9 Marriott, Chapter 8; Jenkins, Chapter 12; Elkin, Chapter 4.
10 Elkin, 26.
11 Elkin, 58–9.
12 Elkin, 54–6.
13 Marriott, 43.
14 Bronwen Edwards and David Gilbert, '"Piazzadilly!" The Re-Imagining of Piccadilly Circus (1957–72)', *Planning Perspectives* 23, no. 4 (2008): 463.
15 Jenkins, 224.
16 Ministry of Transport and Civil Aviation, *Report of the Committee on London Roads (The Nugent Report)* (London: HMSO, 1959), 18–19.
17 Marriott, 138; Edwards and Gilbert.
18 Anon., *The Recurrent Crisis of London: CIS Anti-Report on the Property Developers* (London: Counter Information Services, [1973]), 8.
19 Edwards and Gilbert.
20 *Guardian*, 20 December 2015, www.theguardian.com/business/2015/dec/20/harry-hyams; Marriott, 141.
21 Donald Foley, *Controlling London's Growth: Planning the Great Wen 1940–1960* (Berkeley: University of California Press, 1963), 126.
22 Jenkins, 223.
23 Unnamed source, quoted in Elkin, 73.
24 Marriott, 140.
25 Jenkins, 221.
26 David Starkie, *The Motorway Age: Road and Traffic Policies in Post-War Britain* (Oxford: Pergamon Press, 1982), 28.
27 Anon., *The Recurrent Crisis*, 13.
28 Corporation of London Improvements and Town Planning Committee, *The City of London: A Record of Destruction and Survival* (London: Architectural Press for the Corporation of London, 1951), 284.
29 Simon Bradley and Nikolaus Pevsner, *The Buildings of England. London 1: The City of London* (London: Yale University Press, 2002), 130–3.
30 *Guardian*, 8 July 1959, 3.
31 Bradley and Pevsner, *The City of London*, 540–5; see also David Heathcote, *Barbican: Penthouse over the City* (Chichester: Wiley-Academy, 2004).
32 Bradley and Pevsner, *The City of London*, 542.
33 Films included *Life at the Top, Blow Up, Don't Raise the Bridge, Lower the River, Sebastian*, and *Crossplot*. Simon James, *London Film Location Guide* (London: Batsford, 2007), 199–200.
34 Peter Hall, *London 2000* (London: Faber and Faber, 1963), 213.
35 Edward Carter, *The Future of London* (Harmondsworth: Penguin, 1962), 158.
36 Reyner Banham, *The Architecture of the Well-Tempered Environment* (London: Architectural Press, 1969).

37 BBC, *A City Crowned with Green* (film), 1964.
38 Michael Hebbert, 'The City of London Walkway Experiment', *Journal of the American Planning Association* 59, no. 4 (1993): 433–50.
39 Hebbert, 'The City of London Walkway Experiment', 440.
40 Hebbert, 'The City of London Walkway Experiment', 434.
41 Marriott, 87–91.
42 Jenkins, 224.
43 Hebbert, 'The City of London Walkway Experiment', 437.
44 Hebbert, 'The City of London Walkway Experiment', 444.
45 Anon., *An Aid to Pedestrian Movement: A Report by a Working Party on the Introduction of a New Mode of Transport in Central London* (London: Westminster City Council, 1971); see also Edwards and Gilbert.
46 Anon., 'City Plan Failure?', *The Pedestrian: Quarterly Journal of the Pedestrians' Association for Road Safety* 62 (Autumn 1966): 22.
47 Hebbert, 'The City of London Walkway Experiment', 445.
48 Heathcote, 223.
49 Heathcote, 224.
50 Alan Powers, 'The Heroic Period of Conservation', *Twentieth Century Architecture* 7 (2004): 8–18; Michael Hebbert, *London: More by Fortune than Design* (Chichester: John Wiley & Sons, 1998), 85–6; Jenkins, 256–7.
51 Hebbert, 'The City of London Walkway Experiment', 446.
52 Bradley and Pevsner, *The City of London*, 144.
53 There is a coda to this story of a comprehensive pedway network, in that small sections of new pedway, realized in architecturally bold weathering steel, started to be built to replace some demolished former sections as part of Oxford Properties and Brookfield's 'London Wall Place' development in 2017. Thanks to Michael Hebbert for alerting me to their comeback.
54 Marriott, 87.
55 Simon Bradley and Nikolaus Pevsner, *The Buildings of England. London 6: Westminster* (London: Yale University Press, 2003), 771–5.
56 Francis Cave, *The Pimlico Precinct Traffic Scheme* (London: City of Westminster City Engineer's Department, 1968).
57 *The Times*, 15 February 1966, 14.
58 Francis Cave, 'Environmental Improvements in Relation to Housing and Traffic', *Royal Society of Health Journal* 90, no. 2 (March 1970): 77.
59 City of Westminster Archives Centre: 'City of Westminster City Engineer's Report to Traffic Committee', by Francis Cave, 15 February 1966, in 'Westminster City Council Traffic Committee Minutes'; 'Westminster City Council Traffic Committee Minutes', 4 April 1967.
60 Cave, 'Environmental Improvements', 78.
61 Cave, *The Pimlico Precinct Traffic Scheme*, 4.
62 Carmen Hass-Klau, *The Pedestrian and City Traffic* (London: Belhaven Press, 1990), Chapter 4.
63 Hass-Klau, Chapter 5.
64 Hass-Klau, Chapter 6.
65 Hass-Klau, Chapter 8.
66 Hass-Klau, 170–1.
67 Hass-Klau, 236.
68 City of Westminster Archives Centre: 'City of Westminster City Engineer's Report to Traffic Committee', by Francis Cave, 15 February 1966, in 'Westminster City Council Traffic Committee Minutes'.
69 Marriott, 110.
70 Douglas Sutherland, *The Landowners* (London: Anthony Blond, 1968), 89.

71 Marriott, 110; *The Times*, 7 July 1965, 6; *Daily Telegraph*, 5 July 2002, 27.
72 Marriott, 108; Jenkins, 218.
73 *Evening News*, 5 July 1971, 6.
74 Jenkins, 217; *Evening News*, 6 July 1971, 6.
75 Anon., *Pimlico: Houses for People not Profit* (London: Pimlico Neighbourhood Aid Centre Housing Group and the Pimlico Tenants and Residents Association, 1973), 3.
76 Jenkins, 220.
77 Donald Appleyard, *Livable Streets* (Berkeley: University of California Press, 1981), 183.
78 Harold Scott, *Scotland Yard* (London: André Deutsch, 1954), 239.
79 Frank Mort, 'Fantasies of Metropolitan Life: Planning London in the 1940s', *Journal of British Studies* 43, no. 1 (January 2004): 144.
80 *Observer Weekend Review*, 11 December 1966, 1.
81 Anthony Sampson, *Anatomy of Britain* (London: Hodder and Stoughton, 1962), 415.

Bibliography

City of Westminster Archives Centre: Westminster City Council Papers.
Hansard.
Daily Telegraph.
Evening News.
Guardian.
Observer.
The Times.
Anon. *An Aid to Pedestrian Movement: A Report by a Working Party on the Introduction of a New Mode of Transport in Central London.* London: Westminster City Council, 1971.
———. 'City Plan Failure?'. *The Pedestrian: Quarterly Journal of the Pedestrians' Association for Road Safety* 62 (Autumn 1966): 22.
———. *Pimlico: Houses for People not Profit.* London: Pimlico Neighbourhood Aid Centre Housing Group and the Pimlico Tenants and Residents Association, 1973.
———. *The Recurrent Crisis of London: CIS Anti-Report on the Property Developers.* London: Counter Information Services, [1973].
Appleyard, Donald. *Livable Streets.* Berkeley: University of California Press, 1981.
Banham, Reyner. *The Architecture of the Well-Tempered Environment.* London: Architectural Press, 1969.
BBC. *A City Crowned with Green* (film), 1964.
Bradley, Simon, and Nikolaus Pevsner. *The Buildings of England. London 1: The City of London.* London: Yale University Press, 2002.
———. *The Buildings of England. London 6: Westminster.* London: Yale University Press, 2003.
Carter, Edward. *The Future of London.* Harmondsworth: Penguin, 1962.
Cave, Francis. 'Environmental Improvements in Relation to Housing and Traffic'. *Royal Society of Health Journal* 90, no. 2 (March 1970): 75–9, 86.
———. *The Pimlico Precinct Traffic Scheme.* London: City of Westminster City Engineer's Department, 1968.
Corporation of London Improvements and Town Planning Committee. *The City of London: A Record of Destruction and Survival.* London: Architectural Press for the Corporation of London, 1951.
Edwards, Bronwen, and David Gilbert. '"Piazzadilly!" The Re-Imagining of Piccadilly Circus (1957–72)'. *Planning Perspectives* 23, no. 4 (2008): 455–78.

Elkin, Stephen. *Politics and Land Use Planning*. Cambridge: Cambridge University Press, 1974.

Foley, Donald. *Controlling London's Growth: Planning the Great Wen 1940–1960*. Berkeley: University of California Press, 1963.

Hall, Peter. *London 2000*. London: Faber and Faber, 1963.

Hass-Klau, Carmen. *The Pedestrian and City Traffic*. London: Belhaven Press, 1990.

Heathcote, David. *Barbican: Penthouse over the City*. Chichester: Wiley-Academy, 2004.

Hebbert, Michael. 'The City of London Walkway Experiment'. *Journal of the American Planning Association* 59, no. 4 (1993): 433–50.

———. *London: More by Fortune than Design*. Chichester: John Wiley & Sons, 1998.

James, Simon. *London Film Location Guide*. London: Batsford, 2007.

Jenkins, Simon. *Landlords to London: The Story of a Capital and Its Growth*. London: Constable, 1975.

Marriott, Oliver. *The Property Boom*. London: Pan, 1967.

Ministry of Transport. *London Traffic Congestion: Report by the London and Home Counties Traffic Advisory Committee*. London: HMSO, 1951.

Ministry of Transport and Civil Aviation. *Report of the Committee on London Roads (The Nugent Report)*. London: HMSO, 1959.

Mort, Frank. 'Fantasies of Metropolitan Life: Planning London in the 1940s'. *Journal of British Studies* 43, no. 1 (January 2004): 120–51, 167–72.

Powers, Alan. 'The Heroic Period of Conservation'. *Twentieth Century Architecture* 7 (2004): 8–18.

Sampson, Anthony. *Anatomy of Britain*. London: Hodder and Stoughton, 1962.

Scott, Harold. *Scotland Yard*. London: André Deutsch, 1954.

Starkie, David. *The Motorway Age: Road and Traffic Policies in Post-War Britain*. Oxford: Pergamon Press, 1982.

Sutherland, Douglas. *The Landowners*. London: Anthony Blond, 1968.

8 Conclusions and epilogue

Introduction

This book has attempted to deconstruct the traffic problem, treating traffic as a socio-technical network of actors, both human and non-human, situated in sociological and political contexts with histories. To do this, the canonical plans and planners have been decentred, allowing an exploration of the margins. This has meant examining conceptualizations, rather than measurements, of congestion, recognizing that it must be seen as more than a case of speed, volume, and flow. It might be possible to conclude from this that we will never be able to understand the traffic problem, as it is just too slippery in our hands. What is seen as a problem from different perspectives seems soluble, but when we fragment things and look at it from the different perspectives considered in this book, it becomes so multi-dimensional that it resembles a kaleidoscope, an ever-changing resolution, interesting in each configuration but never stable. However, we can draw a series of wider conclusions from the patterns, even as they shift and reform in our hands. This concluding chapter expresses themes and connections that have emerged from this approach and its examination of the case studies.

The next section in the chapter connects aspects of the individual case studies together, demonstrating that we can only understand the traffic problem by looking across time, space, disciplines, and professions. It is easy to take the traffic problem at face value, seeing it in turns as a chronic pathology of free movement and a looming acute crisis requiring drastic action. A survey of the characterizations of traffic congestion in London's urban planning literature found a great deal of consistency in the ways it has been represented and in the solutions put forward – circulation, flow, ring roads, segregation, arterial routes, and so on. The traffic problem was a failure of planning.

But in his 1964 BBC film on urban planning, the architectural critic Reyner Banham offered a different view:

> For most of this century, planners and visionaries have denounced London as congested, inhuman, restless, wasteful. But the margin between fullness and congestion is very slight. Often it is only in the eye of the beholder. London

can be like a party where there are just a couple of guests too many ... But most of the time ... a party that really swings.[1]

It has been shown that the vast diversity of experiences of the capital's streets cannot be reduced to a singular, universal problem that only reconstruction can solve. The dominance of professional planning in twentieth-century discourses of traffic and movement in cities has occluded and marginalized wider views and alternative conceptualizations, but we can only understand traffic by seeking and exploring them.

The second section of the chapter draws together ideas about the relationships among traffic, capital, markets, and the state. It proposes that traffic and road infrastructure have a distinct relationship with capitalism, and that London's traffic landscape has a longer history of marketization than is commonly proposed, not just in the overtly neoliberal road pricing project discussed in Chapter 5 but in the saturation with market decisions of a wide variety of traffic projects realized from the 1950s onwards. It is therefore proposed that thinking of traffic as a flat network need not imply that all actors have equal power – capital acts as an ineluctable gravitational pull on all decisions about traffic. However, it is not the only form of social relation to do so.

The final section considers how much of what has been explored is particular to London and how much could apply elsewhere. It will be argued that London's physical and historical geometry renders its traffic unique, but with powerful connections to other places that will give the present account value as a model for thinking about complex urban infrastructures and their problems more generally.

There then follows an epilogue that concludes the book with a brief survey of the traffic problem in London today using the case studies as our guide, demonstrating the long lives of infrastructures and the power they wield in shaping our conceptions of the urban experience.

Deconstructing 'the traffic problem'

The first research question addressed in this book asked what have been the dominant characterizations of the traffic problem, and what alternative characterizations have been obscured by the dominant framings. The evidence presented in the exploration of the planning literature in Chapter 2 shows how the urban planning discourse proposed physical reshaping of London with a new road network based on circulation, fast flows, and segregated access, and that this discourse dominated thinking about traffic problems and congestion throughout much of the twentieth century. The traffic problem was a failure of planning and, closely allied, a failure to build. However, this was not the only conceptualization of a traffic problem. The case studies examined in Chapters 3 to 6 show a series of alternative views, from the problem being a machine failure, to it being a failure to control people, a failure of the market, or a system failure. Chapter 7 proposed that the traffic problem is intimately linked with the accumulation of capital, whether overt or otherwise, and can be seen therefore in many senses as a manufactured problem in which solutions created markets and opportunities for profit – more on this presently.

By treating traffic as a socio-technical network it has been possible to reveal these marginalized accounts and place them on an even footing with the dominant account of reconstructionist planning. At a micro-geographical level this leads us to see a city of adjacencies – a calm residential neighbourhood in Pimlico a few yards away from the main thoroughfare of Victoria Street, for instance, or the East India Dock Road turning into the Victoria Dock Road. As has been demonstrated in the case studies, both the characterization of the traffic problem and the solutions enacted differed greatly as one moved a few paces from one place to the next – traffic calming to tower-block building in one instance, guard rails to an elevated concrete flyover in the other. We can only discern the fine grain of the problem by breaking it down. Nevertheless, focusing solely on the local and the microscopic risks short-sightedness. So, we need at the same time to take a macro-geographical approach, looking at people, ideas, and technologies on global scales, and we will return to this in a moment. History, too, has opened the black box of the traffic problem to scrutiny. By taking the problem at face value out of history, out of time, we treat it as constant and unchanging, and more worryingly, we risk treating and judging the actions of the past on our own terms, rather than those contemporary with the events being studied. Congestion in the same place in 1970 and 1980, for instance – or even 1972 and 1974, either side of the collapse of the Ringways – meant different things each time.

Practices of marginalization

Given the significance of these alternative characterizations of traffic, we have benefited from being able to develop an understanding of the processes of marginalization. Chapter 3 looked at the construction of the Silvertown Viaduct in east London, a location at the literal margins of the capital, straddling the LCC area and the County Borough of West Ham. It was also a marginal location of the British Empire, a site of transit and transformation between the metropolis and the periphery, mediated by the ships and lorries transporting goods between the margins and the centre. The concrete viaduct, its construction partly enabled because it was located out of sight, far from metropolitan aesthetic concerns, became therefore an imaginary structure expressing an engineering world view, with concrete and grade separation lifting citizens up above the old tangled and inefficient road network and offering a new, efficient perspective on the Empire – the view of the docks from above – as well as a new conception of empire – a Labourite democracy open to all, contrasted with a Fascist autocracy elsewhere in Europe.

Chapter 4 looked at the guard rails erected along the East India Dock Road. This was a project that sought to place pedestrians in the literal margins of the road, segregating road users into categories, casting some as 'impure' or 'abject', and discriminating over their place in the mobility infrastructure of the capital. On the one hand, this conceptualization of the traffic problem as a failure of control can be seen as a contrast with the democratic view expressed in the previous chapter. Yet the two projects were contiguous and formed a single traffic route, from the docks to the centre of London and vice versa. So too the world views can be seen as connected, as grade separation – the flyover – lost its 1930s democratic inclusiveness to become the dominant segregationist technology absorbed by the planning discipline and expressed from the 1950s onwards.

Chapter 5, which explored road pricing and its later manifestation as the congestion charge, described in detail the active marginalization of the economic voice by Colin Buchanan's planning team. This led to many years of lobbying by those involved in the neoliberal project to move the discourse of marketization of road space from the political margins to the mainstream, a process that in one sense failed, but that, by taking a longer view and understanding the complex relationships between neoliberal ideas and social democracy, can easily be seen to have taken root, its political stock rising as that of the reconstructionist planners fell in the 1970s and 1980s. In the neoliberal economy, price is the means by which segregation is really enacted, rather than steel rails or concrete flyovers.

Chapter 6 examined electronic networks of traffic signals, vehicle sensors, and surveillance cameras in an early manifestation of today's notions of smart cities. Here, the conceptualization of the traffic problem was of system failure, the inability of older methods of traffic control to make the most of an existing physical infrastructure. This was expressed in clear opposition to the reconstructionist paradigm but, unlike the economists, could work equally well if new roads were built – it simply necessitated a reprogramming of the system.

Why is this characterization of the traffic problem sitting in the margins? It sits at the margins of our vision; in other words it is hiding in plain sight. The sensors of this electronic traffic network are buried just an inch below the tarmac. Its switches hide in plain black metal boxes planted in the pavement – boxes we never notice. Traffic lights are all but invisible owing to their ubiquity. And the traffic cameras over our heads, connected to sophisticated computer networks, have multiplied and grown like Topsy to an extent few realize. The location of the computer rooms that run this network are secret, and the control centres are open only to those with high security clearance. The market model of resource allocation relies on knowledge of the state of the market at any instant. The electronic network now installed across London's streets is providing that knowledge in finer and finer detail. This may yet become the ultimate expression of the neoliberal state and we are barely aware of it.

Friction, not congestion

Those, then, are the marginal stories brought to the centre. Now, having examined a range of characterizations of congestion as the traffic problem, we can conclude with one that emerges from the mobilities turn in the humanities, which is to characterize congestion not as a *problem* but as *friction*. To frame congestion as a problem, one must believe that it can be solved, however difficult or expensive the solution may be. Such a characterization sets congestion as *other than* mobility. Yet this does not hold up under scrutiny. Reyner Banham was hinting at this in his comments mentioned earlier. The hard-line conclusion might be that there is no such *thing* as congestion. This is a useful corrective position, and certainly appealing to the relativist mindset, but perhaps goes a little too far to be especially illuminating.

A more moderate position would be to consider congestion (however one wishes to define it) as *constitutive of* mobility – in other words, it is an inevitable experience of being mobile. Peter Adey has commented that the flows of mobilities are not smooth but disrupted, often in political ways – the security queues at airports are an example.[2] This is friction. Tim Cresswell has commented in a similar fashion on friction as 'the stubborn stickiness of what is often called "the real world"'.[3] He has defined friction in the sense of impediment to motion, but also as a means of enabling it.[4] He reacts against critics of the mobilities turn who argue that its focus on flows 'turns our attention away from the stickiness of space and place', as he terms it. He argues instead that:

> an approach to the world that foregrounds mobilities should, instead, provide a position that highlights friction ... Sometimes it slows and stops the mobility of people, things and ideas, and sometimes it enables them. Mobility is often impossible without friction.[5]

This helps explain the Silvertown Way, which offered a new view over the docks, a sticky site where the products, ships, and citizens of empire came temporarily together before flowing away in all directions again. The flyover smoothed the flow but was also designed to allow people to gather and gaze. Friction also helps explain the guard rails at the East India Dock Road, which ended up doing nothing to prevent lateral flows across the street: shopkeepers had the keys to gates in the rails, people ducked under and over the rails, and life continued to happen *in* the street as well as in places connected *by* the street. They rubbed along together. In this characterization, there is no such thing as free-flow, just frictional flow. This, then, is our more moderate position. Rather than claiming that there is no such thing as congestion, we can say that *all* movement is congested.

Viewing congestion as friction, and thus essential to movement, is a shift in the way we consider traffic, and raises the stakes somewhat. It is an argument that congestion is not a by-product, but *the* product, and all studies of traffic mobilities must inherently be studies of traffic congestion. Congestion is not a failed state, but rather it is the normal state, a *required* state, and should thus be treated not as a problem requiring solution but as a state requiring understanding. We could go further. Taking into account Deirdre Boden and Harvey Molotch's argument that human societies have a 'compulsion of proximity', even with advanced technologies that might dilute needs for co-presence, we might conclude that congestion does not just constitute *mobilities*, it constitutes *cities*. Boden and Molotch argue that 'Copresent interaction remains ... *the* fundamental mode of human intercourse and socialization.'[6] Therefore it is logically impossible to have a city without congestion, because congestion *forms* cities.

Traffic infrastructure and capital

The second research question asked about the relationships between London's traffic infrastructure and capital. Traffic congestion is about people as much as vehicles and roads, and characterizations of the traffic problem, and solutions put forward to solve them, shed important light on the politics of mobilities. Decentring the heroic planning story and focusing on people and ideas at the fringes of professional urban planning has thus allowed us to see beyond the canonical plans, as well as seeing those plans as unstable artefacts. Yet a flat network need not imply a flat ontology. We returned to the planning profession in Chapter 7, examining ways in which aspects of the canonical plans were negotiated into reality. In doing so we have been able to consider the complex relationships among capital, planning, and mobilities, helping us understand some of the motivations behind the characterizations people have constructed of traffic and congestion.

Here, we can conclude that those involved with the accumulation of capital have marginalized *themselves* from the traffic problem – because by doing so they have been able to take advantage of the opportunities it has created to extract profit. Some of these have been obvious – the ultra-secretive Harry Hyams hiding behind his architect, Richard Seifert, in the building of Centre Point. Other opportunities have been more surprising – at the Pimlico Precinct, congestion was engineered *into* residential streets, at least for drivers wishing to pass through, increasing amenity and opportunities for gentrification. This work was done by Westminster Council planning officials and (allegedly) a parade of property developers whose identities cannot be recovered owing to the byzantine complexity of their company structures. Marginal figures one and all. Profit can be extracted whatever the traffic problem looks like, and it can often be hard to follow the money.

The exchange value of infrastructure

We can find answers to this question of capital in broadly Marxian accounts of place-making and cities as spaces *of* capital. David Harvey has explored the ways places have been 'erected as permanences within the flux and flow of capital circulation'.[7] Such studies come down, at a micro scale, to accounts of property, specifically buildings. But the spaces in between – traffic infrastructure – have been intensely capitalized too, and they bring their own opportunities to make profit as well as feeding into efforts to profit from conventional property. They are not gaps but built structures in their own right. Thus, roads, flyovers, pavements, and precincts can all be read in this mode, particularly in terms of

Harvey's comments on the 'strongly differentiated' landscape that results, and 'the rise of spatially ordered (often segregated) social distinctions'. Infrastructure has exchange value. The tensions of capitalism discussed in this book between, say, Centre Point and London Wall (the problem was congestion, leading to an opportunity to profit) and the Pimlico Precinct (the problem was traffic flowing too freely leading to an opportunity to profit) can be explained as part of the 'speculative element' in such schemes.

There is much to learn from this approach. The 'speculative element', claims Harvey, means that whilst capitalism is involved in making places, this 'is not to argue that the geographical pattern is determined in advance'. This all helps to explain the apparent failure of planning to solve the traffic problem. Top-down holistic planning would deny capital's opportunities to profit through speculation and to pit one opportunity against another – competition. Harvey goes on to discuss creative destruction – spatial reorganization – and its role in enabling further capital accumulation. His account places the surge of such reorganization from around the year 1970, which fits with conventional chronologies of neoliberal activity. Yet in the case studies examined here, we have seen evidence of capitalistic place-making and creative destruction rather earlier – certainly back to the 1950s and strongly evident in the 1960s. The streets of London have been 'saturated with market decisions' (in David Gilbert's words) throughout the period of the postwar consensus as well as after its collapse.[8]

What this all points to is the absurdity of the 'economic cost of congestion' trope, in which economists, business leaders, motorist organizations, and the national press claim repeatedly that London's traffic jams cost the economy so many billions of pounds per year. By looking differently at the situation, congestion can be clearly seen as a crucial part of the money-*making* system, albeit in ways that can be unexpected or apparently contradictory.

Capitalism or neoliberalism?

In many respects, this is an age-old story of laissez-faire capitalism. Property speculation off the back of infrastructure improvements is, of course, at least as old as industrialization – with Metroland and the ribbon development of the arterial roads just two examples from London's history. Left alone, capitalists will speculate to accumulate, and capital is therefore present in every urban process. But there is a new story here too, which is to do with the extent of the state's involvement in this activity, and the setting up of structures (of governance or planning) that enable marketization and capital accumulation. By looking askance at the postwar planning of London – conventionally a top-down story of holistic, technocratic state intervention and the power of the planner – we can sketch out a revised periodization of the history of neoliberalism, and a new way to see capital in the traffic landscape.

There are three aspects to this revision. The first is the example of road pricing told in Chapter 5. If the surge of neoliberalization of the capital's economy has usually been placed in the 1970s, the road pricing story has pulled that periodization back to the late 1950s and 1960s with the Smeed Committee and the work of the IEA. This was neoliberalism in opposition to state planning. The second aspect comes from other accounts considered here – Centre Point, London Wall, and the Pimlico Precinct. It would be wrong to see neoliberalism as an ideologically pure doctrine of small-statism. Philip Mirowski has suggested that neoliberals, amongst themselves, 'would admit that their agenda was to take over existing governments, and maybe even make them larger, more intrusive, and more powerful, as long as they managed to implement explicitly neoliberal policies'.[9] The three 1950s/1960s episodes discussed in Chapter 7 have shown the intimacy (albeit messy and manipulative and taking different forms) between property speculators and council planning departments. This was something different from laissez-faire capitalism. The state did not leave the capitalists well alone, as that description might suggest. In the neoliberal reading, the state and the speculators manipulated each other in order to advance their goals in a process that saw the marketization of infrastructure.

The third aspect of the new periodization of London's neoliberal story, albeit one that might only be coming to maturity today, concerns the 1960s traffic sensor and surveillance network described in Chapter 6. If, as Mirowski has argued, neoliberalism expands the classical liberal tradition to 'encompass the whole of human existence as a political animal and a knowing being', then knowledge of the market is the crucial currency of the movement. Mirowski has claimed that 'the market was deemed a greater information processor than any human being and the only reliable arbiter of truth', so there is distinct value in viewing *real* information processors – such as the computer-controlled electronic network under and over the capital's streets – in this light.[10] Now vehicle sensors and plain CCTV cameras have been joined by automatic number-plate recognition cameras, face recognition cameras, and audio-recording devices, and these are increasingly connected to databases – of banking details, of vehicle identities, of individuals, of conversations, of telephone calls, and many other vast sets of data.[11] Opportunities for capital accumulation arise from knowledge and predictions of the flows of the market. Here, the vehicle sensors and computer modelling can be seen as providing that, and thus might become vital tools of the neoliberal project, if we accept that the neoliberals don't really care about small states and *your* individual freedom so long as they are in control.

Agency and power in infrastructure

Yet, this is not just about capital, which might be the most obvious and powerful form of power in the history of congestion but not the only one in town. We wish to understand more widely how power intersects with disaggregated assemblages. We have therefore examined episodes in characterizing the traffic problem where the hand of capital seems, if not absent, at least detached. Alker Tripp's concerns were about categorization and control inflected more by a social and body politics than an overtly economic one. Thus, his segregation was a moral crusade and needs to be understood in a wider frame. Similarly, the world views of engineers Henry Maybury and Charles Bressey were formulated through ideas of efficiency, flow, and circulation, where efficiency related to governance as well as traffic routes (Tripp held this view too). Thinking of the traffic problem as a set of networks of interrelationships, with no hierarchical precedence of people or artefacts, raises interesting challenges about the nature of traffic, and considers how artefacts as well as human actors have agency.

Put simply, is the traffic problem socially constructed or technologically determined? As has been suggested, the answer must be neither, or rather both. Congestion is itself a network characteristic, and the case studies have demonstrated time after time that the ways people see, characterize, and attempt to solve the traffic problem are inflected by both social contexts and the affordances of technologies: novel ones, and ones already embedded in the urban landscape. Reinforced concrete gave engineers a technology with which to imagine new roads soaring over old ones in more rational formations. But ideas about replacing cluttered street layouts with more geometrically efficient ones dated back at least to Christopher Wren. Silvertown Way was built just as much out of a desire to open up working-class areas with wide, straight roads as it was the consequence of the invention of reinforced concrete. Traffic signals; guard rails; and, later, electronic computers, vehicle sensors, and CCTV cameras gave police officers and traffic controllers a technology with which they could imagine bringing the users of streets to order. But guard rails along the East India Dock Road were installed just as much because of 1930s ideologies of segregation and prohibition. Only after Vietnam War demonstrations did CCTV cameras flourish. The technologies were *necessary* but not *sufficient* to explain why the police wanted to keep people in their place. Similarly, developments in metering technologies enabled both neoliberals and social democrats to construct traffic problems, but the political and economic ideas behind their characterizations – market failure and environmental damage – were much bigger than the technology.

We must therefore conclude that issues such as pedestrian segregation tell us as much about race, gender, class, and the effects of colonialism as they do about capital, and the ideas put forward in Chapter 4 of this book – the racist guard rails and promiscuous pedestrians – would tend to reinforce that view. Capital strongly affects the shape of the network of mobilities but is not the only form of power relation that does so. What has been described here is a historical geography of human agency and the particularities of places, albeit operating within political systems broadly construed. It is messier than any simple account could describe, and it is argued that its consequences are for the most part unintended. It has been constructed by actors over long periods and in many places taking advantage of opportunities presenting themselves in particular circumstances. Specific local conditions have affected the construction and materialization of congestion, as well as the global ideas we have considered here. At the risk of sounding like a stuck record, there is no single traffic problem, so there can be no single account of congestion. What has been proposed here is a set of wider concerns that a study of congestion can shed some light on.

London – the unique city?

The third research question asked what is distinctively 'London' about London's traffic problem, and what the relationship is between London and other places. The first conclusion is rather banal, and is to say that there *is* something distinctive about the traffic problem in London as opposed to other places. The Danish planner Steen Eiler Rasmussen certainly believed London was a unique city, as he lobbied for the rejection of 'all pattern-made recipes for town planning' in favour of what Frank Mort has described as London's 'innately democratic and localized use of space'.[12] Time and again, discussions about the problem have focused on ideas of grid layouts, and attempts made either to rebuild London with such a layout of streets or somehow to impose onto its existing layout grid-like behaviour. This has included planning visions such as those of Le Corbusier and the MARS group, realized in small part as the London Wall development. It has included visions of control, with Alker Tripp's proposed grid-like temporal pulses of traffic along highways. The engineers such as Charles Bressey tried literally to overlay a grid of concrete streets over the existing pattern. The scientists building an electronic traffic city had to experiment in grid-plan Glasgow to make sense of their models. The Pimlico Precinct, on the other hand, was set up to de-grid the behaviour of a localized grid neighbourhood. All this alongside the constant refrain about Christopher Wren's lost opportunity.

It must be concluded that the congestion in London is formally different from that in New York or Glasgow, because the street pattern differs so formally. It has fewer parallel routing options. It must also be concluded that the river matters: for all sorts of historical and topographical reasons the Blackwall Tunnels are a nightmare. In a geographical study, a conclusion that physical geography affects mobility is inevitable. There are other particularities of London that have been observed repeatedly in this account, with its patchwork multi-tier governance structure perhaps the most significant. London is a unique product of such structures, and the particular people who have acted within them.

On the other hand, a second banal conclusion is that there are aspects of this story that would apply equally to other places, however one wishes to compare and contrast. Every time we try to focus on one place – London, in this instance – we find ourselves making mental jumps to other places to see how their cultures have influenced London, and how London has influenced them. This is what happens when one thinks in terms of networks. Time and again in the account presented here, ideas developed elsewhere have found their way to London, helping shape the traffic landscape, as well as making their way out to other places. We have recovered detailed stories of the technologies, debates, practices, and implementations constituting those networks – the unexpected connections. We have glimpsed an international circulating network of people and ideas about traffic. These ideas have been applied (and of course modified and appropriated) in a wide variety of cities, whether they are planning visions moving among English Garden Cities, American Beautiful Cities or French Radiant Cities; policing visions moving between the Chicago Police Department and Scotland Yard; economic visions circulating among the universities of Iowa, Yale, Birmingham, and Cambridge; or visions of electronic surveillance and control in Washington DC, Los Angeles, Glasgow, or Hammersmith. There has been a globalization of conceptualizations of traffic problems through the twentieth century.

From 'the traffic problem' to 'world city traffic'

But, beyond these straightforward or banal conclusions, there is a more significant one if we consider this globalization from the point of view of world cities.[13] The work of Doreen Massey offers important insights here, and connects with ideas expressed earlier about seeing congestion alternatively as friction. The first point to make is that whilst there may have been a globalization of approaches to traffic, this does not equate to homogenization, nor does it mean people with more mobility must be richer than those with less. The experience of mobilities in cities is by no means common from city to city or person to person. This draws on Massey's ideas of 'power geometry' explored in her seminal work looking at places such as Kilburn High Road. 'Different social groups', she claims,

> have distinct relationships to this anyway differentiated mobility: some people are more in charge of it than others; some initiate flows and movement, others don't; some are more on the receiving-end of it than others; some are effectively imprisoned by it.[14]

The conclusion must be that actors in the network of traffic are unevenly placed with respect to congestion (or friction, in the new characterization) and the experience depends on how *in charge* the actor is. Massey's argument is that we need to consider the specificities of places, as well as the social power-relations emerging out of places, to understand stories of globalization. She observes that 'a local–global politics would be structured differently from place to place'.[15] In this mode, each actor in their own context constructs a different traffic problem. This local construction of congestion points up the value in studying micro-geographies of traffic, but in open-ended ways.

This, then, is the final conclusion of this book. The 1940s image of the Tottenham Court Road, captioned 'The traffic problem' in the Penguin version of *The County of London Plan*, and set into historical context in Chapter 2, is repeated in Figure 8.1.[16] Let us examine more closely what it is showing us.

Figure 8.1 'The traffic problem', *County of London Plan*, 1943–1945 (London County
Council).

Firstly, the motor-cars to the bottom-left do not appear to be in the queue for the
junction. They are taxi cabs waiting at the centre of the road, the position where all
West End taxi ranks were located at that time.[17] The cabbie standing chatting by his
vehicle might have been somebody like Herbert Hodge, who had lived and worked
in North America as a lumberjack, farmhand, political activist, playwright, and
hobo before returning to his native London to drive a taxi (and narrate Charles
Bressey's 1939 film *The City*, discussed in Chapter 3).[18] Or the taxi driver in the
picture might have been somebody like Maurice Levinson, born in Russia, who had
been forced to leave aged six months following an anti-Semitic pogrom. His family
travelled as refugees across Europe before settling in Stepney.[19] These taxi drivers,
depicted in a photograph entitled 'The traffic problem', hardly express ideas of
stasis. They had travelled the world – albeit with varying control over their
journeys – and settled in London for the opportunities it brought. Taxi driving
in the Tottenham Court Road, for them, was about freedom, not imprisonment.
Perhaps the taxi driver in our photograph was discussing that.

Or let us look at the buses waiting to cross over the junction into Charing
Cross Road. The number 24 bus route is one of the oldest in London. We can

just make out on its destination blind some of the places it passed through on its journey that day. It started at Hampstead before reaching Chalk Farm and went on through Camden Town to Euston before reaching Tottenham Court Road. After that it ended up travelling down Whitehall on its way to Westminster, before reaching Victoria and finally Lupus Street towards its terminus at Pimlico.

It passed by two major teaching hospitals, the Royal Free and University College Hospital, and would have carried the lower-paid hospital staff to and from work – nurses and orderlies from around the country and the Empire who had ended up in the great imperial metropolis. When the picture was taken in 1943, the foundation of the NHS, and the great voyage of the *Windrush*, were just five years away. The 24 bus was part of a global network of mobility. Perhaps its passengers were talking about job opportunities in the capital. Maybe some of them were returning to their rented rooms in Pimlico houses north of Lupus Street, little suspecting that seven years later the duke of Westminster would sell the estate and begin a process of developer-led gentrification that would see them driven from their homes.

The bus also carried civil servants and countless thousands of government functionaries and officials to and from Whitehall and Westminster as they prosecuted the business of a world city and an imperial nation. It carried factory workers, too. Camden Town was home to industry and manufacture, including global brands such as Gilbey's gin (now part of Diageo) and Idris soft drinks (now part of Britvic). The 24 bus carried their employees to work, and into the West End for play.

We can focus on the rear bus. It shows an advertisement for Craven Plain cigarettes, made by Carreras, a firm founded in Somers Town in the nineteenth century by an immigrant Spanish family. The area had seen a huge influx of Spanish refugees like them in the 1820s, all fleeing political unrest. In 1928 (and then owned by Bernhard Baron, a Russian Jewish refugee) the firm had opened a huge new tobacco factory in Mornington Crescent (also on the 24 bus route) that employed new air-conditioning technologies imported from America and a concrete construction system brought over from France. It was built in the Egyptian style, influenced by the opening in 1922 of Tutankhamun's tomb, and was a famous landmark on the Hampstead Road, passed and marvelled at by passengers on the 24 bus as well as being visited by architecture and engineering specialists from around the world. The factory was opened by Leslie Hore-Belisha, who was to become transport minister six years later. Its 400,000 square feet of maple flooring were imported from Canada, where Carreras had a factory.[20] Asphalt for the external landscaping was mined from Pitch Lake in Trinidad, owned by a British company.[21]

This could go on and on. For instance, we could see where each vehicle in the photograph was manufactured, and think about the automotive industry that was scaling up production to meet wartime needs and an expected postwar consumer and export boom that would rely on generating demand for driving.[22] We could look at the theatre at the edge of the picture, and consider the centrality of London's theatre-land to the global circulation of plays and movies. Most of London's theatres were in the middle of changing ownership from actor-managers to financial speculators, who sought profit and needed crowds.[23] We could think about the life journey of everybody in this picture – there would be hundreds, if we saw past the vehicles.

The point is simply that. We should see beyond the gathering of *vehicles*, and see this in an open way – see it inside out, as Massey has said. This picture captured a moment of friction in a world city – a moment when countless global human mobilities came together in a bit of stickiness amid the energy and restlessness of the capital. A moment that happened because of London's standing in the world – because of what London *is*. A moment expressing the complex power geometries of globalization.

Why have we not solved London's traffic problem? This book has shown that it is a multi-dimensional set of constructions, rather than a singular problem, and therefore cannot be treated in isolation from wider concerns. Moreover, it is shot through with capital and other forms of power, creating advantages as well as disadvantages. Congestion in a world city – world-city traffic – is inextricably intertwined with globalization, and it is a safe bet that those who benefit from globalization will always find ways to benefit from London's traffic.

Epilogue: London's traffic problem today and tomorrow

It is January 2016. In a building 100 yards from the traffic crowding up Tottenham Court Road, an exhibition opens to the public entitled *Streets Ahead: The Future of London's Roads*. It has been curated by New London Architecture (NLA), a non-profit information and discussion organization chaired by architect Peter Murray that seeks to 'bring people together to shape a better city' through debating architecture, planning, development, and construction.[24]

The exhibition is laid out like a network diagram, with coloured lines on the floor (akin to those on the tube map) guiding visitors from one exhibit to the next. Unlike the tube network, though, there is only one linear route through the exhibition, which begins with ancient history – the failure of London planners to solve the problems of the streets, from the London Society to Colin Buchanan and the Ringways – and ends with futuristic visions of autonomous vehicles and freight deliveries by drones. Opening the exhibition is the claim that 'Working well, streets and roads keep London thriving, but when they are clogged up and overcrowded, the city's health, economy and environment suffer.'[25] A few paces on, visitors are told that 'Congestion is undoubtedly one of the biggest challenges, as Londoners already spend 70 hours on average in traffic jams every year', and that 'Too much congestion can damage economic competitiveness.' Later in the exhibition, figures are deployed: 'The cost of congestion is forecast to increase by up to an additional £800 million per annum, on top of the present £5 billion.'

Streets Ahead presents a version of the holistic, networked view of traffic explored in this book. It concludes that 'There is no one solution which tackles all of the challenges.' There are multiple voices here, reflecting the diverse world views examined in the case studies, as opposed to a singular statement expressing 'the traffic problem'. Marie-Claude Hemming of the Civil Engineering Contractors Association comments that 'Providing London's road users with a network fit to keep London moving is a continuous task for civil engineering contractors.' David Leam of London First, representing London's business leaders, claims that 'Without radical interventions such as more sophisticated congestion charging, journey times and reliability on London's roads will only get worse.' London's then-mayor, Boris Johnson, explains that 'we are doing our utmost to cut congestion and delays through the use of innovative new traffic light technology', and the exhibition reveals plans for 'smart systems' that combine 'live data on road use with a predictive signalling system'. Soon, visitors are told, the surveillance systems and sensors embedded underneath and watching over the streets will be joined by mobile sensors fitted to every London bus, which will 'act as "intelligent" sensors: to pinpoint congestion, to measure the condition of the road … and to learn more about passenger movements'. And NLA chair Peter Murray recognizes a shift in ideas about segregation on the streets, commenting that:

> There is a growing realisation that we need to change the hierarchy of our streets – while 20th-century policies focused on the speed of vehicular movement, today we place more significance on health and wellbeing, on safety, and on walking, cycling and public transport – while still permitting the wheels of commerce to function efficiently.

London's traffic authorities have been systematically removing pedestrian guard rails since the mid-2000s.[26]

It is common, in public exhibitions about technology, to present 'the past' and then to express the present as a watershed moment. Here, then, we hear that 'The scale of London's growth and its associated challenges means that once-in-a-lifetime changes and investment in how our roads work need to be made.' But, as this book has shown, the proposals for addressing 'how our roads work', ranging from construction to charging, and from planning to smart sensing, have long histories. Indeed, it is possible to see the ghosts of old schemes creeping back in. The Ringways, the exhibition claims, were abandoned because of 'local campaigns and spiralling cost' – no sense of national outrage and a political and cultural shift to the environment in this assessment. Two-thirds of the way round the exhibition, softened up by streetscape improvements and cycle schemes, visitors are told that 'we need to investigate ways of replacing or creating new space over the coming decades'. The solution is new tunnels, flyunders, and decking over main traffic routes. We come back to the Ringways, then, but under rather than over the capital – just as Forshaw and Abercrombie originally wanted. One visitor was transported back to the past, commenting on a handwritten card that 'this exhibition has been dug out of the 1960s'. They could equally have said the 1940s or even earlier.

Above all this can be read as an exhibition about profit. It has been curated by a consortium of industry professionals rather than the lone academic voice of a Patrick Abercrombie or a Colin Buchanan. The characterization of traffic expressed here is one of opportunities, not problems. Road improvements, the exhibition's curators claim, 'can act as catalysts for wider regeneration, inward investment and local economic growth'. Boris Johnson comments on plans to build traffic tunnels across the city's central areas, claiming that they 'could bring income and investment to those areas'. On work to unravel one-way systems such as that at Tottenham Court Road, as well as other 1960s traffic projects such as gyratories and roundabouts, the exhibition observes that 'These are now being removed or altered to create public spaces, many part-funded by developers and through commercial support.' Big datasets will be connected together that will 'drive efficiencies in resources and budgets', according to John McCarthy of transport consultants Atkins. The commercial drone market for freight delivery 'is predicted by some to grow by up to £7 billion in the next decade', the exhibition claims. This characterization of London's traffic is not a failure of planning, or of control, or of markets or systems. It is a failure to seize opportunities to make money from traffic.

Notes

1 BBC, *A City Crowned with Green* (film), 1964. Quotation starts at 17:51.
2 Peter Adey, 'Mobilities: Politics, Practices, Places', in *Introducing Human Geographies*, 3rd edn, ed. Paul Cloke, Philip Crang, and Mark Goodwin (Abingdon: Routledge, 2014), 794.
3 Tim Cresswell, 'Friction', in *The Routledge Handbook of Mobilities*, ed. Peter Adey, David Bissell, Kevin Hannam, Peter Merriman, and Mimi Sheller (Abingdon: Routledge, 2014), 107.
4 Cresswell, 108–9.
5 Cresswell, 114.
6 Deirdre Boden and Harvey Molotch, 'The Compulsion of Proximity', in *NowHere: Space, Time and Modernity*, ed. Roger Friedland and Deirdre Boden (Berkeley: University of California Press, 1994), 258.
7 David Harvey, 'From Space to Place and Back Again', in *Justice, Nature and the Geography of Difference* (Oxford: Blackwell, 1996), 295–6.
8 David Gilbert, in conversation, 9 September 2015.
9 Philip Mirowski, *How Did the Neoliberals Pull It Off?* (Public Books, 2013), public-books.org/nonfiction/how-did-the-neoliberals-pull-it-off.
10 On recent developments in the system of cameras, sensors, lights, and computers in London's streets, and the increasing ease with which agencies can get access to data, see anon., 'CCTV', *How It Works* 43 (January 2013): 52–7.
11 On early facial recognition technologies, and Home Office interest in their development, see BSSRS Technology of Political Control Group with RAMPET, *TechnoCop: New Police Technologies* (London: Free Association, 1985), 42–52; and David Rooney, *Mathematics: How It Shaped Our World* (London: Scala Arts & Heritage, 2016), 64–9.
12 Frank Mort, 'Fantasies of Metropolitan Life: Planning London in the 1940s', *Journal of British Studies* 43, no. 1 (January 2004): 131. He quotes Rasmussen on same page; for the original source of the quote see Steen Eiler Rasmussen, *London: The Unique City*, rev. edn (London: Jonathan Cape, 1948), 434–5.

13 For a view on London as a world city, see Peter Hall, *London Voices, London Lives: Tales from a Working Capital* (Bristol: Policy Press, 2007), Chapter 1.
14 Doreen Massey, 'A Global Sense of Place', in *Space, Place, and Gender* (Minneapolis: University of Minnesota Press, 1994), 149.
15 Doreen Massey, *For Space* (London: Sage, 2005), 103.
16 E. J. Carter and Ernö Goldfinger, *The County of London Plan: Explained by E. J. Carter and Ernö Goldfinger* (London: Penguin, 1945), 9; originally published (without title and with slightly wider view) in John Forshaw and Patrick Abercrombie, *County of London Plan* (London: London County Council/Macmillan, 1943), facing 8.
17 Herbert Hodge, *It's Draughty in Front: The Autobiography of a London Taxidriver* (London: Michael Joseph, 1938), 182.
18 Jonathon Green, 'Herbert Hodge: The Cabbie Philosopher', *The Dabbler* (blog), 2 May 2013, http://thedabbler.co.uk/2013/05/herbert-hodge-the-cabbie-philosopher/; GPO Film Unit, *The City: A Film Talk by Sir Charles Bressey* (film), 1939.
19 Maurice Levinson, *The Taxi Game* (London: Peter Davies, 1973); John Davis, 'Levinson, Maurice (1911–1984)', *Oxford Dictionary of National Biography*, May 2013.
20 David Hayes, 'Carreras: Family, Firm and Factory', *Camden History Review* 27 (2003): 32.
21 Anon., 'Notes on Illustrations: The New "Carreras Building"', *Architecture* 7, no. 38 (May 1929): 83.
22 William Plowden, *The Motor Car and Politics in Britain 1896–1970* (London: Bodley Head, 1971), Chapter 15.
23 Maggie Gale, *West End Women: Women and the London Stage 1918–1962* (London: Routledge, 1996), 39–43.
24 New London Architecture website, www.newlondonarchitecture.org/about.
25 This and subsequent quotations from the exhibition are taken from New London Architecture, *Streets Ahead: The Future of London's Roads* (exhibition catalogue) (London: New London Architecture, 2016).
26 *The Times*, 8 April 2006, 38.

Bibliography

New London Architecture website, www.newlondonarchitecture.org/about.
The Times.
Adey, Peter. 'Mobilities: Politics, Practices, Places'. In *Introducing Human Geographies*, 3rd edn, ed. Paul Cloke, Philip Crang, and Mark Goodwin, 791–805. Abingdon: Routledge, 2014.
Anon. 'CCTV'. *How It Works* 43 (January 2013): 52–7.
———. 'Notes on Illustrations: The New "Carreras Building"'. *Architecture* 7, no. 38 (May 1929): 83–4.
BBC. *A City Crowned with Green* (film), 1964.
Boden, Deirdre, and Harvey Molotch. 'The Compulsion of Proximity'. In *NowHere: Space, Time and Modernity*, ed. Roger Friedland and Deirdre Boden, 257–86. Berkeley: University of California Press, 1994.
BSSRS Technology of Political Control Group with RAMPET. *TechnoCop: New Police Technologies*. London: Free Association, 1985.
Carter, E. J., and Ernö Goldfinger. *The County of London Plan: Explained by E. J. Carter and Ernö Goldfinger*. London: Penguin, 1945.

Cresswell, Tim. 'Friction'. In *The Routledge Handbook of Mobilities*, ed. Peter Adey, David Bissell, Kevin Hannam, Peter Merriman, and Mimi Sheller, 107–15. Abingdon: Routledge, 2014.

Davis, John. 'Levinson, Maurice (1911–1984)'. *Oxford Dictionary of National Biography*, May 2013.

Forshaw, John, and Patrick Abercrombie. *County of London Plan*. London: London County Council/Macmillan, 1943.

Gale, Maggie. *West End Women: Women and the London Stage 1918–1962*. London: Routledge, 1996.

GPO Film Unit. *The City: A Film Talk by Sir Charles Bressey* (film), 1939.

Green, Jonathon. 'Herbert Hodge: The Cabbie Philosopher'. *The Dabbler* (blog), 2 May 2013. http://thedabbler.co.uk/2013/05/herbert-hodge-the-cabbie-philosopher/.

Hall, Peter. *London Voices, London Lives: Tales from a Working Capital*. Bristol: Policy Press, 2007.

Harvey, David. 'From Space to Place and Back Again'. In *Justice, Nature and the Geography of Difference*, 291–326. Oxford: Blackwell, 1996.

Hayes, David. 'Carreras: Family, Firm and Factory'. *Camden History Review* 27 (2003): 30–6.

Hodge, Herbert. *It's Draughty in Front: The Autobiography of a London Taxidriver*. London: Michael Joseph, 1938.

Levinson, Maurice. *The Taxi Game*. London: Peter Davies, 1973.

Massey, Doreen. *For Space*. London: Sage, 2005.

———. 'A Global Sense of Place'. In *Space, Place, and Gender*, 146–56. Minneapolis: University of Minnesota Press, 1994.

Mirowski, Philip. *How Did the Neoliberals Pull It Off?* Public Books, 2013. publicbooks.org/nonfiction/how-did-the-neoliberals-pull-it-off.

Mort, Frank. 'Fantasies of Metropolitan Life: Planning London in the 1940s'. *Journal of British Studies* 43, no. 1 (January 2004): 120–51, 167–72.

New London Architecture. *Streets Ahead: The Future of London's Roads* (exhibition catalogue). London: New London Architecture, 2016.

Plowden, William. *The Motor Car and Politics in Britain 1896–1970*. London: Bodley Head, 1971.

Rasmussen, Steen Eiler. *London: The Unique City*, rev. edn. London: Jonathan Cape, 1948.

Rooney, David. *Mathematics: How It Shaped Our World*. London: Scala Arts & Heritage, 2016.

Index

Printed and bound by CPI Group (UK) Ltd, Croydon, CR0 4YY

24/10/2024

01778282-0009